Oil and Gas Trading

A Practical Guide

Consulting Editor **Denys Hickey**

Consulting editor
Denys Hickey

Managing director
Sian O'Neill

Oil and Gas Trading: A Practical Guide
is published by
Globe Law and Business Limited
3 Mylor Close
Horsell
Woking
Surrey GU21 4DD
Tel: +44 20 3745 4770
www.globelawandbusiness.com

Oil and Gas Trading: A Practical Guide
ISBN 9781911078036

Table of contents

Introduction

Denys Hickey
39 Essex Chambers

The initial idea for this book came from the oil trading cases I have handled since the late 1970s. Over the intervening period, the oil trading market has undergone many changes, with the rise in the importance of independent traders following the first oil crisis in 1973. The *'Seven Sisters'*, as the major oil companies were known, have gradually lost power and the independent traders have gained power along with national oil companies. Add to this the entry and, more recently, the exit of the big commercial banks from the market, and the rise in importance of derivatives, and you have a market which has seen more change than any other with which I have been involved.

Against this background, it seems to me that there was a need for a general guide to the practical aspects of oil trading. Although the underlying law relating to oil trading is based on standard sale of goods and shipping law, there are numerous special features of the oil trading market that make it unique. This short guide does not pretend to be a detailed textbook, but aims to introduce the general reader to some of the special features of the trade. If any reader wishes to do more research, then I would recommend *Benjamin's Sale of Goods, Scrutton on Charterparties, Chitty on Contracts* and the various practitioners' textbooks on trade finance.

One of the peculiarities of the trade is that there is no one set of standard terms and conditions, although both the oil majors and the national oil companies have their own general terms and conditions (GT&Cs) and the independent traders seek to impose GT&Cs of their own. The fact remains, however, that there is no homogenous set of terms that govern all transactions. Because of this, it is important to understand the basic principles that apply to the trade, whether it involves crude oil, refined oil products or gas.

The separate chapters of this book have been written by individuals with enormous experience of the areas in which they have contributed, from firms including Minton, Treharne & Davies Ltd, who are the worlds foremost experts on oil measurement and quality issues with crude and refined products.

We also have contributions from Stuart Shepherd on delivery and nomination and from Jane Fitzgerald of Transammonia on the determination of quantity and quality. Peters & Peters, who have an enormous amount of experience of dealing with sanctions and trade embargoes, have contributed both on sanctions and on joint selling. There is a very useful guide to derivatives and hedging from Catherine Jago, and a very clear guide to trade finance written by Anna Koshy and Simon Cook of Sullivan & Worcester UK LLP.

Cheryl Guan and John Simpson of Ince & Co Singapore have written on oil cargo insurance. Paul Aston of Holman Fenwick & Willan Singapore has written chapters on bills of lading and the main features of trading contracts, on the basis of many years' experience of handling trading disputes. Iain Sharpe of Rodyk Singapore has contributed on the very important subject of damages. Last, but not least, Ben Olbourne, a colleague of mine who is now based in Singapore, has written about the carriage of oil.

The recent fluctuations in the price of crude oil underlines the importance of understanding the basics for anybody involved in oil trading. As we all know, oil prices have fluctuated wildly over the last few years, which demonstrates that oil is an extremely volatile commodity in which massive amounts of money are at stake. The rights and wrongs of disputes can depend in some cases on very small variations in contract terms or in the way situations are handled by traders and claims handlers.

I trust that this book will go some way towards shining a light on the principles that apply to trading contracts in the oil market, and will allow the reader to navigate safely between and around the obstacles.

Formation and conclusion of international oil and gas trading contracts

Denys Hickey
39 Essex Chambers

This chapter discusses what needs to be covered in order to create an enforceable contract and what terms may be implied into the contract. The author also comments on the application of general terms and conditions (GT&Cs), to which, in his experience, traders do not always pay sufficient attention.

1. Introduction

Independent oil traders are a relatively new phenomenon, having only emerged in the 1970s, following the first oil crisis in 1973. Before that date, the movement of oil tended to be controlled by the big international oil companies, the so-called 'Seven Sisters', and the oil price was low, which did not encourage international trading. Crude oil was largely transported by the oil majors in tankers owned by them, and was refined in their refineries, with the refined products then being distributed through their own petrol station chains. All of that changed with the first oil crisis, when the price rose, the national oil companies took a much more prominent role and the old trading patterns began to change.

The oil trading market now features oil majors, national oil companies, independent oil traders and some financial players such as the big banks, which moved into physical oil trading a few years ago, although they now appear to be withdrawing from the market because of capital and regulatory requirements as well as risk assessment.

This background partly explains the reason for the diverse nature of the physical oil trading market and the fact that no standard contract terms have been developed. Instead, all of the principal national oil companies have their own terms and conditions, as do the oil majors, such as BP and Shell. The big independent traders, such as Vitol, Glencore, Trafigura, Gunvor and Mercuria, also often seek to impose their own terms. All of this is, however, very different to some of the other commodity trades. For example, most sugar is traded internationally on the terms of the Refined Sugar Association, while most grain is traded on the GAFTA standard contract terms. There is, therefore, very little room for confusion when seeking to agree the terms of an individual trading contract in these trades.

This may be contrasted with the oil and gas market, in which the trader who is doing the negotiating is only concerned about price, quality and delivery dates. Such traders are not interested in the detailed terms that go into making up a complete contract, including terms concerning shipment, payment of demurrage or any of the common more technical clauses, because the oil is almost invariably being moved by ship, which means that there are insurance implications and chartering terms to consider.

Against this background, it is useful to consider the basic position and some of the traps for the unwary so far as concluding an oil trading contract is concerned. The following discussion assumes that the intention is for the contract to be governed by English law, as this is the law most commonly used for international oil trading contracts.

2. Concluding the contract – basic rules

The basic rule under English law is easy to state: in order to decide whether a contract has been concluded, a court or arbitrator needs to see an offer, an acceptance of the offer, and evidence that all of the relevant terms have been agreed. What often happens, however, is that one party makes an offer and his counterparty, in responding, will agree to the basic terms but will also seek to impose his own GT&Cs or will seek to modify the terms proposed for the contract. In the author's experience, this often leads to difficulties and disputes.

It may be useful to contrast the position so far as oil trading is concerned with the negotiation of a voyage charter, where there are almost always two brokers involved. The practice is for one of the brokers involved to send out a 'recap' of the terms and ask for confirmation that the recap is agreed. Very often, the response which comes back will seek to modify certain terms and it is only when all of the details have been agreed, and any 'subjects' (such as receipt of management approval or a stem of cargo being available) have been lifted that there is a firm fixture. At that point, the brokers will send out a final recap setting out the agreed terms. Accordingly, as a matter of evidence, it is usually very clear whether or not a charter has been concluded.

The challenge with oil trading contracts is to determine whether an agreement has been concluded. This is because the negotiations very often take place on the telephone and may be partly recorded on email. In addition, over the last few years negotiations have also increasingly been conducted using instant messaging applications, by virtue of which the traders concerned have a number of windows open at the same time and discuss their requirements backwards and forwards with other traders and brokers. Under English law, all of this material is potentially relevant to determining whether or not an agreement has been concluded. The difficulties that can be encountered are demonstrated by two cases in which the author was involved and which are considered below.

2.1 The *Vitol* case

In the first case, *Vitol SA v Deepak Fertilisers*[1], Vitol (for which the author acted), sold a cargo of 3,000 metric tonnes of methanol to Deepak for delivery in India. The parties had previously entered into some 29 contracts for the sale of methanol over

1 Jonathan Hirst QC, sitting as a Deputy Judge of the High Court, judgment dated May 21 2008, unreported.

an eight-year period, for quantities totalling some 79,000 metric tonnes. This trading ceased in 2007 because of a difference between the parties over demurrage, which Deepak was reluctant to pay. However, the parties discussed resuming their trading relationship and agreed to enter into some spot contracts.

Against this background, it is useful to set out the details of the negotiations because they illustrate the difficulties that can occur. The negotiations started in January 2007 with a telephone conversation between the parties which was immediately followed by an email confirming the sale of 3,000 tonnes of methanol. A so-called 'indent' was then sent, which the parties referred to as the 'written agreement' and which was in the same terms as had been used in previous contracts. After the indent was signed, cargo was shipped to Deepak in January 2007 and paid for.

On February 1 2007, Deepak contacted Vitol's trader and said that it was interested in buying 3,000 metric tonnes of methanol for delivery in the second half of March. In an email sent later in the day, Vitol proposed offering the quantity requested for delivery in March, with the price to be based on an average price formula. Immediately after the conversation, Deepak emailed to Vitol as follows: "Okay, accepted by Deepak". The brokers who were involved also sent an email to Deepak concerning the sale. On the next day, the brokers duly emailed Deepak with a copy of the sale 'indent', together with some details concerning the wording of the required letter of credit. They ended their message by asking Deepak to return the sale indent "duly stamped and signed". None of these email exchanges resulted in Deepak reverting to Vitol denying that there was a contract.

Nothing more was heard until February 15 2007, when Vitol nominated a vessel and asked Deepak to confirm its acceptance. About a week later, Deepak confirmed that the vessel was accepted and asked for the indent to be forwarded.

At about this time, it became clear that the price of methanol was in sharp decline and Deepak sought an amendment to the pricing regime. This was not accepted by Vitol and it was only then that Deepak, for the first time, queried whether a contract had in fact been concluded. The basis of Deepak's refusal to acknowledge that there was a concluded contract was that it had not signed the contract or a letter of intent. Deepak developed this argument as the case progressed, saying that it was common practice in the industry to treat business as only having been concluded once written terms were signed off on the terms of the sales indent.

In the event, the judge decided that a contract had been concluded on February 1 2007 on the basis that the market was one in which deals were made by telephone. He went on to comment that he did not understand how business could be transacted if the contract were not to be treated as concluded for some considerable time afterwards on the basis that it had to await signature of a formal written document. He also mentioned the obvious practical point that it was only when the market turned against it that Deepak sought to argue that there was no concluded contract.

---

2 [2008] 2 Lloyd's Rep 685.

2.2 The *Statoil* case

The second case is *Statoil SA v Louis Dreyfus Energy Services LP*[2] (in which the author was involved on behalf of Statoil, the well-known Norwegian state-owned oil company). Statoil claimed a total of $435,833 in respect of demurrage due under the terms of a contract for the sale of LPG made in 2006. The background was that Statoil concluded a charterparty for the LPG carrier Harriette N for one voyage from Norway to several possible discharge ports, with Louis Dreyfus to nominate its final choice in due course. The negotiations for this contract were conducted through a broker, and the two traders at Statoil and Louis Dreyfus respectively did not email one another directly or talk to each other on the telephone. All material communications were therefore made by the respective traders emailing the sole broker, and there was only one relevant telephone call between the trader at Statoil and the broker.

On August 2 2006, a Mr Anderson of Statoil sent an email giving a "firm indication" for the sale of 32,000 tonnes of LPG, CIF Mexico including laytime of 30 days and a demurrage rate of $40,000 a day or *pro rata*. Later that day, the broker forwarded a firm counter-indication from Louis Dreyfus, with the only difference being the price. On the following day, August 3 2006, a response was forwarded by the broker from Statoil, with the only difference again being the price. Then, later in the afternoon of that day, the trader at Louis Dreyfus sent an email to the brokers with a counter-offer, setting out the full terms of a proposed contract and, under the heading 'Demurrage', a clause as follows:

> *As per charter party*
> *Any demurrage claim shall be submitted by seller within 90 days after completion of discharge. Any claim made after this time bar shall not be valid. Buyer and seller shall agree to the demurrage amount by 90 days from buyer's receipt of seller's claim. Thereafter, if any dispute outstanding, buyer shall pay all undisputed amounts promptly.*

In addition, it was stated toward the end of the email that SITCO GT&Cs should cover the transaction and that "seller's contract administration will issue a full sales contract upon fixing". In other words, assuming the contract was agreed, it would be Statoil, through its Contract Administration Department, that would be responsible for finalising and issuing the full contract terms, which the judge found was usual in the trade.

Later still in the afternoon of August 3 2006, the broker sent an email to Louis Dreyfus recording that the counter-offer was accepted, except for a number of matters, and crucially, at the end of the email, stating that "Statoil general terms and conditions" were to apply. Louis Dreyfus accepted this position in an email but asked that the Statoil standard terms and conditions should be sent to Louis Dreyfus for its review.

Finally, a recap of what purported to be the full terms of the contract was sent out and, under the heading 'Demurrage', recorded the demurrage rate of $40,000 per day or *pro rata* and stated that any demurrage claim should be submitted within 90 days after completion of discharge.

On the next day, August 4 2006, the broker sent Louis Dreyfus an email with a link to Statoil's GT&Cs and, in the meantime, the trader at Statoil decided that he wanted to use the wording of a recent contract that had been concluded between the parties as the template for the contract terms. The evidence which was accepted by the judge was that the Statoil trader had telephoned the broker to explain this, and had heard the broker talking to the trader at Louis Dreyfus at the same time, whereas the Louis Dreyfus trader denied having such a conversation, and his denial was accepted by the judge.

At the conclusion of these negotiations, there was a situation in which there had been no response to the last email sent by the trader from Louis Dreyfus, and Statoil did not issue the contract on Monday, August 7 2006. The evidence was that the Statoil trader had created a 'deal sheet' on Statoil's internal computer system but did not ask Statoil's Contract Administration Department to prepare the contract terms until the next day. The contract terms recorded the demurrage rate as $40,000 per day and incorporated Statoil's GT&Cs, and were only typed up by the Statoil Contract Administration Department on around August 11, but for some reason were not sent to Louis Dreyfus until September 20 2006.

The vessel went on to deliver her cargo that September and, following discharge, Statoil sent invoices for demurrage totalling $774,357 to Louis Dreyfus. The main issue to be determined by the judge was what were the terms of the contract? In particular, did they contain a demurrage time bar?

The principles applied by the judge on the basis of *Pagnan SPA v Feed Products*,[3] were that the court had to review what the parties said and did and infer their objective intentions. The relevant material comprised the email exchanges between Statoil and the brokers and between the brokers and Louis Dreyfus. It is therefore useful to look at the detailed findings of the judge because similar situations often arise in connection with negotiations for oil trading contracts.

It was clear that Statoil's first proposal did not include a demurrage time bar, whereas the response did include one. The response via the brokers said "thanks your counter to which Statoil accept/except ..." It was held that in responding in this way, Statoil had indicated that it rejected the demurrage time bar. However, this case provides a classic illustration of how careful traders and their contract administration teams need to be concerning the finalisation of contract terms. To illustrate further the problems that can occur, it is worth pointing out that the contract recap produced by Statoil and sent only on September 20 2006 did not contain any time bar clause. Discussions on the terms were, however, still going on as late as September 25.

The lesson from this case is that if a party wishes to avoid what turned into a two-day trial, with all the accompanying expense, it is wise to make sure that as soon as negotiations are concluded that the terms agreed are set out in writing at the earliest opportunity rather than being left unrecorded.

3. Is agreement on all terms needed?

It can happen that although there is broad agreement on contract terms, not all of the details have been discussed and agreed. In that situation, is there a concluded contract?

---

3 [1987] 2 Lloyd's Rep 615.

The answer to this is that it depends. Courts and arbitrators will seek to enforce a contract where all of the essential terms have been agreed, but will not rewrite a contract if the parties have failed to agree all of the essential terms. Care must therefore be taken to include all of the essential clauses in the contract. The question which a court or arbitrator will ask is whether or not the parties reached agreement on all of the essential terms. If they did not do so, then the court or arbitrator will not rewrite the contract for them – although, if it is possible to interpret the terms that have been agreed so as to complete any gaps, then courts and arbitrators are likely to do that, provided that it is clear that the parties intended to conclude a contract. Courts and arbitrators will not, however, make up wording if a gap has been left by the parties; the result may then be a finding that no contract was concluded.

The other aspect is that, although an agreement not to negotiate with anyone else for a limited period can be enforceable in circumstances such as a bid, an agreement to negotiate will normally be unenforceable, on the basis that it is merely an agreement to agree. Parties should therefore be careful to make sure that negotiations result in a definite agreement, rather than a mere agreement to negotiate further. A good example of the problems this can cause is where parties conclude an annual supply agreement with the prices fixed for the first year, but with prices to be negotiated for subsequent years. If no agreement can be reached on prices for the second year, then the question is whether courts or arbitrators can impose a price in the absence of agreement between the parties, an issue that has arisen in relation to a number of long-term supply contracts.

The answer under English law is that a provision for the parties to agree upon a new price for years subsequent to the first year of a supply agreement will be unenforceable. This is because such a provision is a mere agreement to agree, and such a bargain is something that is not certain enough to be enforceable by courts or arbitrators.

Accordingly, the common sort of clause whereby the parties agree a firm price for first year of the contract and agree to negotiate so far as the price for subsequent years is concerned will be unenforceable unless a mechanism is put in place according to which the new price can be determined. As always with contracts, everything will depend upon the wording of the clause, but for the purposes of this chapter, the only important things to remember are that unless there is a clear price review mechanism setting out the criteria to be applied in order to determine any price change, a court or arbitrator will not be in a position to decide upon a new price to apply for the period under review.

The way to get round this if it is desired to build some flexibility into the pricing under a long-term supply contract is to set out specific criteria or formulae that can be applied in the event that the parties fail to agree on a revised price. The difficulties which this can bring about, however, can be illustrated by the recent fall in the price of crude and the knock-on effect on other fuels. In many cases, a supplier will not wish to allow any revision in the price if, for example, the supply contract forms part of a network of agreements that have been used to justify the development of an offshore field, where often the sale of the associated oil and gas may be concluded on a 'take or pay' basis. In these circumstances, the owners of the field, as sellers of the resulting oil and gas, will not want to have any variation in the price obtained for the resulting crude oil or gas because it is likely to underpin the financing of the field in question.

There have, over recent years, been a number of price review arbitrations that illustrate how difficult this area can be and why it is essential to have very clearly drafted provisions if the parties wish to have the option of adjusting the price in the light of events as the contract progresses.

The lesson from the cases is that parties should make sure that all of the essential terms are agreed and are drafted in clear language that does not leave any room for doubt about their intentions.

3.1 Formalities

Under English law, commercial contracts generally do not have to comply with any particular formalities, apart from specialised contracts such contracts for the sale of land and some consumer contracts. Oil trading contracts do not have to follow a particular form, and can be concluded by an exchange of emails, faxes or telexes, or as a result of a series of telephone negotiations, provided that it can be proved that there has been an offer and an acceptance, and sufficient terms have been agreed to allow the contract to operate.

This is very different to the position under some other legal systems, where a contract will not be valid unless the terms are set out in writing and are physically signed by both parties. As anyone involved in the oil trading world knows, this would make the conclusion of many oil trading contracts extremely difficult, bearing in mind that many such contracts are made between parties located in different jurisdictions.

One particular area in which there have been developments in the last 20 years is in relation to contracts concluded by means of electronic communications. Article 9(1) of the European Directive on Electronic Commerce 2000 provides that:

> *Member States shall ensure that their legal system allows contracts to be concluded by electronic means. Member States shall in particular ensure that the legal requirements applicable to the contractual process neither create obstacles for the use of electronic contracts nor result in such contracts being deprived of legal effectiveness and validity on account of their being made by electronic means.*

One of the main issues is whether or not this directive requires contracts to be evidenced in 'writing'. The Law Commission advised the UK Government that the requirements for contracts to be in 'writing' and the requirement for a 'signature' could be fulfilled via electronic means so that email would suffice, provided that it can be proved that the relevant emails have been sent and received. This directive was implemented by the Electronic Communications Act 2000, which provides for the admissibility in legal proceedings of electronic signatures and related digital signatures (and effectively restates the existing position under English law).

3.2 Authority

One area which, in the author's experience, can sometimes cause difficulties is where one of the parties to the contract alleges that the person who agreed the contract did not have authority from the company to do so. This has been the case in particular with jurisdictions where managers may not be authorised to bind their company to any substantial commitment.

In some jurisdictions, it is easy to find out who has authority and the limit of that authority, as such matters are recorded in a public register. In other jurisdictions, however, it is not always easy to be sure who has authority, which can pose real problems for companies seeking to enter into oil trading contracts.

Under English law, the question of whether or not the individual concerned has authority to enter into a contract will depend on English law rules. This will require an examination of all the correspondence and negotiations to determine whether that person had actual authority, or was held out by the company as having authority (referred to as ostensible authority), which may often be the case if there have been previous transactions that have been concluded by that individual and his or her authority has not been questioned.

The more difficult aspect, however, is that when a claimant seeks to enforce a judgment or arbitral award in the defendant's home country, the same defence of lack of authority may be raised again and can be more difficult to overcome. It is suggested that a party should take advice before entering into a contract with a counterparty from a jurisdiction outside the UK, follow whatever guidance is proffered concerning the method of concluding a contract, and check the authority of the individuals concerned. If the contract is significant enough, it may be wise to obtain direct confirmation from the company concerned that the individuals who sign are authorised to do so, by either a properly executed resolution of the company board or a power of attorney executed in accordance with local law in the jurisdiction where the counterparty is based. While this may sound like overkill, in the author's experience it can prevent an awful lot of wasted time and expense. A good illustration of the problems that can occur is provided by *The Jin Man*,[4] in which the South Korean guarantor argued that a guarantee had been was signed by the chief executive officer without the authority of the company because no Board approval had been obtained. While the court decided on a preliminary basis that the claimants had the better argument that the CEO had ostensible authority to sign, the case shows how careful a party needs to be.

Another matter to check is whether the contract needs to be written in the local language in order to be valid. A good example of a country in which this is the case is Indonesia. A failure to comply with such a requirement may result in the inability of a party to enforce any judgment or award.

3.3 Incorporation of terms

Many traders do not appreciate that it is not necessary to incorporate a lengthy set of bespoke terms and conditions in order to conclude a workable contract for the sale of oil or gas. This is because the original Sale of Goods Act 1893 represented a codification of existing case law on the sale of goods; although it has been superseded by the Sale of Goods Act 1979 ('the 1979 Act') and various amendments,

4 [2009] EWHC 2941.

the position remains that, under English law, a number of important terms will be implied into a contract for the sale of oil or gas unless there has been an express agreement to exclude or modify the statutory terms.

Alongside the 1979 Act, the English common law has developed a large body of case law concerning the sale, transportation, financing and insurance of goods that is more than adequate to deal with any situation which may arise. Provided that the main terms regarding dates, price, quality and terms of delivery have been agreed, it is perfectly possible to agree a perfectly workable contract that will be governed by general principles of English law, including the 1979 Act.

The author would also caution that in his experience it is often the case that traders do not understand the terms on which they are contracting. This is partly because many traders assume that all oil company terms are much the same, whereas in fact they can be radically different from each other. A good example is that Shell International (SITCO) used to have GT&Cs that were very different from those of Shell UK (SUKO).

Some GT&Cs are biased in favour of the buyer, some in favour of the seller, and others can be very difficult to apply to real life trading situations because the person drafting the terms may have been a lawyer and wrongly assumed that the people administering the contract would always have the terms by their side and would consult and abide by them. Life is not, however, like that, and the author has often been asked to advise on a contract incorporating terms of which his client did not have a copy. In other words, the client was operating a contract governed by rules about which it was completely ignorant. This is not a recipe for a happy ending.

A good illustration of the difficulties that can occur is the standard terms that were imposed by Enron in its bespoke ISDA Master Agreement relating to oil futures trading. Those terms provided that if Enron had any doubts about the creditworthiness of a counterparty, it could terminate the contract concerned, whereas counterparties did not have the same rights as against Enron. This proved to be extremely important when Enron got into severe financial difficulties, which eventually led to the company becoming insolvent. A large number of oil traders had open positions with Enron that they could not close out because the Enron terms did not allow this, and as a result they had to allow those positions to run to maturity without knowing whether Enron would be able to perform.

Another issue encountered by the author on several occasions is where traders have tried to 'improve' their position without necessarily understanding the consequences. A good example is a case in which the author was involved in which the contract provided for a letter of credit (L/C) to be opened by a certain date and went on to provide that if the L/C were not open by that date, the buyer would pay interest for the time the L/C was outstanding. In these circumstances, the normal rule under English law is that, without this additional provision, the seller could have accepted the failure to open the L/C as a repudiation of the contract and so could have treated it as having been terminated. However, the fact that the parties had provided a remedy for a failure to open the L/C on time meant that the contract could not be terminated and the seller was only entitled to recover interest for the period of delay in opening the L/C. In the particular case where this happened, the result was that the seller lost the right to terminate the contract with a buyer it would have preferred not to continue trading with.

3.4 *Force majeure*

'Force majeure' clauses are commonly included in contracts even though traders are often not aware of their nature. Such clauses are the subject of a common misconception among traders, who normally think that *'force majeure'* is a term of art and will apply without any special term being incorporated. That would be the case under some legal systems. Under English law, however, the term *'force majeure'* has no fixed meaning, and English law only recognises a doctrine of frustration, under which a contract will be treated as terminated if, and only if, it becomes impossible of performance through no fault of either party. English law will enforce a *'force majeure'* clause, but only on the basis that it is to be treated as an exclusion clause and so will be interpreted strictly in accordance with its terms.

Accordingly, if the parties wish to incorporate a *'force majeure'* clause, they need to do so expressly and to think very carefully about its precise ambit. This is because unless such clauses are carefully drafted, the result may be a situation in which the contract is suspended indefinitely; this has happened on several occasions in relation to the trading of North Sea oil when production platforms have been out of action for several months. This can wreak havoc with a trader's position because a trade that was expected to close in, say, January may still be open two, three or four months later and, when the production platform resumes operations, the wording of the *'force majeure'* clause may only have suspended obligations under the contract, so that the seller will be expected to deliver the parcel when production resumes. Equally, the buyer will be expected to take the cargo at the price originally agreed, even though the trading landscape and the market have in that time fundamentally changed. There is no easy solution to this potential problem, but readers are urged to consider carefully what they would like to happen if that sort of situation were to occur and build it in to the wording of their clauses. It is always a good idea to 'stress test' such clauses in order to test what the result would be in the event of a shutdown of production, loss or damage to the producing field/wells/export facilities, or civil unrest or action that directly impacts upon the ability of the seller to deliver and of the buyer to take delivery.

A related issue is the imposition of sanctions, which can be unpredictable and may have very serious consequences, including criminal penalties and the potential invalidity of the underlying contract. One precaution that can be taken is to screen the names of the companies and individuals involved before the contract is signed, and to carry out further checks if there are any circumstances that could give rise to potential sanctions problems. A good illustration of such problems is a trading case in which the author was involved, where the carrying vessel was trading under a name that gave no indication that the ultimate owners were Iranian and were therefore at that time subject to US sanctions. The contract was accordingly terminated by the buyers, but not before potential criminal liability had been incurred in the US.

One suggested precaution worth taking is to make sure that counterparties state in writing that they are not subject to sanctions and are not aware of any circumstances that could involve a breach of sanctions legislation anywhere in the world.

3.5 Exemption clauses

It is very sensible to include an exemption clause in the contract in order to avoid facing potentially ruinous claims for damages if there is a problem with the quality of the cargo supplied, or a delay in delivery leads to potentially very expensive consequences, such as a shutdown of production at a petrochemical complex or factory. Because of this, it is very common for sellers to provide that they will not be responsible for indirect losses and that they do not give any warranty concerning the suitability of refined products for any particular use.

3.6 Competition law

This topic is dealt with in detail in a later chapter,[5] but it is worth mentioning that several cases in which the author has been involved concerned traders, including some working for oil majors, who took part in discussions or took action to exclude other traders from a market, in breach of European Union competition law. Bearing in mind that the potential fines can reach 10% of worldwide turnover, the consequences of this could potentially be very serious.

3.7 Law and dispute resolution

(a) Governing law

It is very important to agree on the substantive law to govern the contract because, if the governing law is not specifically agreed, then under English law the governing law will be that with which the contract has the closest connection, which can be open to dispute. In addition, the Convention on the law applicable to contractual obligations (Rome Convention) 1980 lays down rules for determining the governing law of a contract if the parties have not chosen a governing law. The convention was enacted in the UK by the Contracts (Applicable Law) Act 1990, which entered into force on April 1 1991.

It is therefore important for the parties to set out an express choice of law in the contract. This is because if there is no express choice, a court or arbitrator may decide to apply the law applicable in the home country of a counterparty, or the law applicable at the place of delivery, neither of which may be to the advantage of the other party. As an example, the laws of many European countries provide that exemption clauses are not enforceable unless they are 'fair', which makes the outcome of any dispute difficult to predict. Accordingly, a party must decide whether it wants to choose the law that will apply to and govern the contract, or whether it wants a judge who may have no trading background to do so?

(b) Dispute resolution

A choice should also be made as to how any disputes should be settled. While this may appear rather negative, it is much better to choose either court or arbitration, and also to choose the venue. Arbitration has become increasingly popular because arbitral awards are automatically enforceable in other countries that are parties to the

---

5 See chapter on Joint selling and competition law.

New York Convention on the Recognition and Enforcement of Foreign Arbitral Awards 1958 (the 'New York Convention').

If arbitration rather than the court is selected, the normal choices for oil trading contracts lie between London or Singapore, although there are many other arbitration centres that can be chosen.

The next choice is whether the arbitration should be governed by the rules of an arbitration institution or whether it should be subject to the procedural law of the selected jurisdiction. In England & Wales, this means the Arbitration Act 1996 as amended, or in Singapore, the International Arbitration Act (Cap 143A).

Discussion of the advantages and disadvantages of arbitrating under institutional rules are beyond the scope of this chapter. Broadly speaking, however, under an administered arbitration the arbitral institution has jurisdiction to supervise the arbitration process, whereas with an non-administered (ie, *ad hoc*) arbitration the parties are left to deal with the arbitral tribunal direct, without any interference from the institution.

A common choice in London arbitration is the London Maritime Arbitrators Association (LMAA), although larger term supply contracts are sometimes subject to the arbitration rules of the International Chamber of Commerce (ICC) or the London Court of International Arbitration (LCIA). An important point to bear in mind is that the LMAA does not charge administration fees, whereas the ICC and LCIA do.

In Singapore, the institutional choices are between the Singapore International Arbitration Centre (SIAC), which charges administration fees, and the Singapore Chamber of Maritime Arbitration (SCMA), which does not.

3.8 Implied terms

An implied term is an addition to the express terms of a contract by operation of law. It is important to appreciate that under English law, a contract for the sale of oil or gas will have certain terms implied into it by virtue of the 1979 Act as amended. The most important implied terms under the 1979 Act are as follows.

(a) Description

13. Sale by description E+W+S+N.I.

(1) Where there is a contract for the sale of goods by description, there is an implied term that the goods will correspond with the description.

(1A) As regards England and Wales and Northern Ireland, the term implied by subsection (1) above is a condition.

(2) If the sale is by sample as well as by description it is not sufficient that the bulk of the goods corresponds with the sample if the goods do not also correspond with the description.

(3) A sale of goods is not prevented from being a sale by description by reason only that, being exposed for sale or hire, they are selected by the buyer.

(4) Paragraph 4 of Schedule 1 below applies in relation to a contract made before 18 May 1973.

(5) …

It may seem obvious that the goods supplied must meet their description, but in practice it is not quite so straightforward when dealing with oil cargoes. Take, for example, the expression 'fuel oil', which would seem simple enough. In fact, however, there is no official definition of 'fuel oil' as any oil can in theory be used as 'fuel', although some types will be much better for certain purposes. Even crude oil can be used to power ships' engines after the light ends have been evaporated off. Indeed, the author has dealt with cases in which it was argued that crude oil that had been spilled as a result of the blowing up of the Kuwaiti oil wells by Saddam Hussein's forces met the description of 'fuel oil' because it had been collected in lagoons in the desert and, after a period of some months during which most of the light ends had evaporated, had then been exported and sold as 'fuel oil'. Another example was a series of cases whereby the refined products supplied contained extraneous matter that would have clogged the system if they had been run through a power station or used as bunker fuel for ships. The question there was whether or not fuel oil containing the contaminant still met the description 'fuel oil'.

It therefore pays a party to be extremely careful in describing the product it wishes to buy or sell, in order to avoid any surprises. The tighter the description applied to the crude oil or refined products, the less chance there is of being surprised. Another good example of this is the fact that fuel oil can be 'straight run', or can be produced using a cracking process, in which case it is less suitable for certain types of further processing. A party who wants to be specific should therefore ensure that it makes clear in the contract exactly what type of fuel oil it intends to buy, otherwise the seller will be entitled to rely on the generic description of 'fuel oil'.

(b) Quality or fitness for purpose

Turning to section 14 of the 1979 Act, the implied term concerning quality or fitness for purpose is another provision that has given rise to a number of disputes and reported cases.

14. Implied terms about quality or fitness E+W+S+N.I.

(1) Except as provided by this section and section 15 below and subject to any other enactment, there is no implied term about the quality or fitness for any particular purpose of goods supplied under a contract of sale.

(2) Where the seller sells goods in the course of a business, there is an implied term that the goods supplied under the contract are of satisfactory quality.

(2A) For the purposes of this Act, goods are of satisfactory quality if they meet the standard that a reasonable person would regard as satisfactory, taking account of any description of the goods, the price (if relevant) and all the other relevant circumstances.

(2B) For the purposes of this Act, the quality of goods includes their state and condition and the following (among others) are in appropriate cases aspects of the quality of goods –

(a) fitness for all the purposes for which goods of the kind in question are commonly supplied,

(b) appearance and finish,

(c) freedom from minor defects,

(d) safety, and
(e) durability.

(2C) The term implied by subsection (2) above does not extend to any matter making the quality of goods unsatisfactory –
(a) which is specifically drawn to the buyer's attention before the contract is made,
(b) where the buyer examines the goods before the contract is made, which that examination ought to reveal, or
(c) in the case of a contract for sale by sample, which would have been apparent on a reasonable examination of the sample.

(2D) …

(2E) …

(2F) …

(3) Where the seller sells goods in the course of a business and the buyer, expressly or by implication, makes known –
(a) to the seller, or
(b) where the purchase price or part of it is payable by instalments and the goods were previously sold by a credit-broker to the seller, to that credit-broker, any particular purpose for which the goods are being bought, there is an implied term that the goods supplied under the contract are reasonably fit for that purpose, whether or not that is a purpose for which such goods are commonly supplied, except where the circumstances show that the buyer does not rely, or that it is unreasonable for him to rely, on the skill or judgment of the seller or credit-broker.

(4) An implied term about quality or fitness for a particular purpose may be annexed to a contract of sale by usage.

(5) The preceding provisions of this section apply to a sale by a person who in the course of a business is acting as agent for another as they apply to a sale by a principal in the course of a business, except where that other is not selling in the course of a business and either the buyer knows that fact or reasonable steps are taken to bring it to the notice of the buyer before the contract is made.

(6) As regards England and Wales and Northern Ireland, the terms implied by subsections (2) and (3) above are conditions.

(7) …

(8) …

(9) …

The crucial provision is section 14(2A), which provides that goods will be of 'satisfactory quality' if they meet the standard that "a reasonable person would regard as satisfactory, taking account of any description of the goods, the price (if relevant) and all the other relevant circumstances". Section 14(2B) goes on to provide that, among other things, the following are, in appropriate cases, aspects of quality:

(a) fitness for all the purposes for which goods of the kind in question are commonly supplied.

Accordingly, the seller must do two things:

- keep in mind that section 14(2B)(a) could come into play, and

- unless it is confident that the refined product being supplied is fit for all of the purposes for which it is commonly supplied, restrict the warranty that it is prepared to give.

This is why it is common to see an exclusion of any warranty of fitness in the express terms of the contract.

Main features of contracts for the sale of crude oil and refined products, including LPG

Paul Aston
Holman Fenwick Willan, Singapore

1. Understanding the crude oil market and refined product transacting

1.1 Introduction

Contracts for the sale of crude oil and refined products are ultimately creatures of industry; their terms should therefore not be considered in isolation from the products and markets to which they relate. Given the rich history of these transactions, extensive and comprehensive legal frameworks have developed specifically in relation to these categories of contract. Accordingly, in order to understand the challenges and features of contracts for the sale of crude oil and refined products, it is helpful to provide some background of the scale of the crude oil trade and where the major producers and consumers are based.

1.2 Crude oil products and refining

Crude oil, and its derivative products resulting from refining processes, account for the second largest fossil fuel (after coal) being extracted and produced. Petroleum products are used as fuel for everything, from generation plants producing electricity to bunkers and fuel for ships and motor vehicles, to the manufacture of petrochemicals and liquid petroleum gases (LPG), in addition to numerous other manufacturing processes.

Crude oil and condensate are found in raw natural form, and other petroleum products are derived from refining. The first product out of the refining column is gas, then naptha and progressively heavier component parts of the crude being refined.

Crude oil, fuel oils and some heavy marine diesel oils are referred to as 'dirty petroleum products' (DPP)[1] and most products, from marine diesel to water white jet aviation fuel, are usually referred to as 'clean petroleum products' (CPP).[2]

1 Liquid products refined from crude oil, whose colour is darker than 2.5 on the National Petroleum Association scale. DPPs will often require heating during the voyage, as their viscosity or waxiness would make discharge difficult at ambient temperature. DPPs include fuel oil, low Sulphur waxy residue (LSWR) and carbon black feedstock (CBFS).

2 Liquid products refined from crude oil, the colour of which is less than or equal to 2.5 on the National Petroleum Association scale. CPPs include naphtha, jet fuel, gasoline (petrol), and diesel/gasoil.

1.3 Global oil trade[3]

Global oil trade in 2014 grew by a below average 0.9%, or 490,000 b/d.[4] Import growth was driven by China and emerging economies, while US imports declined due to its increased use of domestic crude in its refineries. Oil exports continued to swing from west to east. China is now the world's largest net oil importer, replacing the United States. China's net imports were just under 7 mb/d,[5] versus 5.1 mb/d in the US.

1.4 Oil reserves

Total world proven oil reserves reached 1700.1 billion barrels at the end of 2014, sufficient to meet 52.5 years of global production. The largest addition to reserves came from Saudi Arabia, which added 1.1 billion barrels. The largest decline came from Russia, where reserves fell by 1.9 billion barrels. OPEC countries continued to hold the majority of the world's reserves, accounting for 71.6% of the global total. Central & South America continue to hold the highest reserves to production (R/P) ratio: more than 100 years. Over the past decade, global proved reserves have increased by 24%, or more than 330 billion barrels.

Total Proved Reserves end 2014

Region	Thousand million tonnes	Thousand million barrels	Share of total
North America	35.3	232.5	13.7%
Central & South America	51.2	330.2	19.4%
Europe & Eurasia	20.9	154.8	9.8%
Middle East	109.7	810.7	47.7%
Africa	17.1	129.2	7.6%
Asia Pacific	5.7	42.7	14.1
World total	239.8	1700.1	100%

3 *BP Statistical Review of World Energy*, June 2015. The statistics in this section come from this source.
4 B/d = barrels per day.
5 Mb/d = million barrels per day.

Total Consumption end 2014

Region	Thousand barrels daily	Share of total
North America[6]	23,347	24.3%
Central & South America	7,125	7.8%
Europe & Eurasia[7]	18,252	20.4
Middle East	8,706	9.3%
Africa	3,800	4.3%
Asia Pacific	30,856	33.9%
World total	92,086	100%

World consumption outpaced global production in 2014, with the former rate increasing by 2.1 mb/d, of which 1.6 mb/d was US increased output. Global consumption increased by 840,000 b/d, with emerging economies accounting for all of the growth. Even though China saw a below-average increase, it still accounted for the majority of the growth in consumption.

Refining: Generally, crude oil is heated and the hot gases which result are passed into the bottom of a distillation column and become cooler as they move up the column. As the gases cool below their boiling point, they condense into liquid. The liquids are then drawn off the distilling column at specific heights, ranging from heavy residues at the bottom, raw diesel fuels in the mid-sections and raw gasoline at the top. These raw fractions are then processed further to make several different finished products.

Although all fractions of petroleum find uses, the greatest demand is for gasoline. One barrel of crude oil contains only 30–40% gasoline. Market demand, however, requires that over 50% of the crude oil barrel be 'converted' into gasoline. To meet this demand, some part of the crude oil must be converted to gasoline. This may be done by 'cracking' (ie, breaking down large molecules of heavy heating oil and residues) or 'reforming' (ie, changing molecular structures of low quality gasoline molecules).[8]

Refining is broken down into four main groups:

- Light Distillates;
- Middle Distillates;

6 Of which the US accounted for 19,035 and 19.9% respectively.

7 Of which the Russian Federation accounted for 3,196 and 3.5% respectively.

8 The next level of complexity adds 'cat cracking and some additional hydrotreating. The most complex refineries add coking, more hydrotreating and hydrocracking.

- Fuel Oil; and
- Others.

The first and the second groups are the main categories of regional consumption by product group.

Product group	Products
Light Distillates	Aviation and motor gasolines, and light distillate feedstock.
Middle Distillates	Jet and heating kerosene, gas and diesel oil, including marine bunkers.
Fuel Oil	Marine bunkers and crude oil used directly as fuel.
Others	Refinery gas, LPG, solvents, petroleum coke, lubricants, bitumen, wax and other refined products.

China and North America consume 10 mb/d of Light Distillates, although North American and global rates of consumption (China excluded) have remained relatively constant since 2004. China's consumption has increased by approximately three mb/d in that period. World consumption was approximately 30 mb/d in 2014.

Middle Distillate consumption in 2014 was 34 mb/d in 2014, up from 29 mb/d in 2004. The main increase in consumption has been in the Middle East and China.

Fuel Oil consumption has decreased from 9.5 mb/d in 2004 to eight mb/d in 2014, again with China and the Middle East being the largest consumers.

The consumption of other products has increased from 18 mb/d in 2004 to 20 mb/d in 2014, with consumption remaining constant worldwide, except for an increase from five mb/d to seven mb/d in China. The rest of the world's consumption has remained constant, with the US as the other large consumer, accounting for about five mb/d throughout this period.

The largest importers of crude oil are Europe, the US, China, India, Japan, and other Asia Pacific nations (in that order). It should, however, be noted that Singapore alone imports 45.6 million tonnes per annum (for refining). Global imports of crude oil amounted to 1876 million tonnes in 2014.

The largest importers of refined products are Europe, the US, Central and South America, other Asia Pacific countries, Singapore, China and Japan. Singapore alone imports 102 million tonnes per annum, but obviously this is not all for domestic use and reflects the fact that Singapore is a centre for the supply of bunkers, as well as providing oil storage and tanking facilities. Global imports of refined products amounted to 911.5 million tonnes in 2014, of which the US, Europe, the former Soviet Union (FSU) and the Middle East were the largest exporters.

Crude oil exports in 2014 amounted to 1876 million tonnes of which the Middle East was the largest, exporting 850 million tonnes, followed by the FSU (294 million tonnes), West Africa (213 million tonnes) and Central and South America (164 million tonnes).

Pricing: Crude oil is priced in US dollars per barrel. It remained well below \$10 p/b[9] until the 1973 oil crisis, after which it reached a level of nearly \$40 in 1980, then falling back to below \$20, mainly until 1999, and hovered at \$60 p/b at the end of 2014, having reached a height of \$120 in early 2011.

Traders normally price their contracts on the basis of premiums or discounts to price indices. For crude oil, these are Brent and WTI (West Texas Intermediate). One of the most well known indices is an index quoted in a publication produced by McGraw-Hill and referred to as 'Platts'. Platts quotes, for example, the Singapore Platts low sulphur fuel oil index ('LSFO') as well as the high sulphur Platts Singapore index ('HSFO'). The HSFO quotation has always been at a discount to the LSFO quotation because of the general availability of high sulphur fuels and the relative higher cost of producing low sulphur fuel oil, either because of the additional processing or because it emanates from a low sulphur (or sweet crude), which is always more expensive than high sulphur crude.

The prices published in their various quotations by Platts are derived from trades reported to the respective officers in charge of their product reporting. Platts quotations in some markets, such as Singapore, are based on market activity in a 'window': this is a 30-minute electronic window during which all bids and offers are recorded. However, Platts LSFO Singapore window has, for example, historically been used very little. This has made it very difficult to report its real value accurately. On the other hand, HSFO Singapore is extremely well traded and the quotation is a truer reflection of the HSFO value in the market. As a consequence of the lack of actual trades in the window, Platts must take into account any offers or bids made in the window. Of course, this opens Platts up to being manipulated either up or down by the participants to suit their own interests. Platts accept that the LSFO is so thinly traded that the price is open to inaccuracies.

A full discussion on the transparency of such indices and their sensitivity to manipulation is not the subject of this chapter, though it has been the subject of discussion in trade articles.[10] The underlying criticism appears to be that the

---

9 P/b = per barrel.

10 On October 9 2012, Bloomberg issued the following report: "Bloomberg – Total SA, France's largest oil company, said it is concerned that inconsistency and a reliance on judgment calls hinders [sic] the accuracy of price reporting agencies in reflecting the state of the oil market. We encounter, several times a year, estimates of market prices on key indices that are out of line with our experience of the day and with the available information on which the price formation is based," Totsa, the Geneva-based trading arm of Total, said in an August 24 letter to the International Organization of Securities Commissions, which was published on IOSCO's website on October 5. The company also said that excluding market participants from the assessment process, or 'boxing,' by price reporting agencies such as Platts, a unit of McGraw-Hill Cos, and by Argus Media Ltd can have "significant economic consequences both on the prices assessed in the market and on the company concerned." The agencies report on various spot and forward prices that are not traded on exchanges. IOSCO was appointed by the Group of 20 nations in November to investigate the role played by price reporting agencies in oil markets. The Madrid-based organization published its final report on October 5 that called for the agencies to adopt 'robust' controls to protect the reliability of the benchmarks. IOSCO stopped short of recommending new rules for traders or trading companies. Madrid-based IOSCO brings together national market regulators from more than 100 countries, including the US Commodity Futures Trading Commission and the UK's Financial Services Authority, to coordinate rules and share information."

information is based on subjective reporting about trades which may or may not be accurate.

2. Overview of oil trading contracts

For a trader, in any one trade there are usually a number of contracts in which he or she is interested:

- the contract for purchase and sale of product;
- a charterparty to carry the oil by sea;
- an insurance contract;
- an inspection contract with local surveyors, to survey in particular the quality and quantity of the oil; and
- a documentary credit for the payment of the price.

There are many different markets in oil trading and to consider each of them is beyond the scope of this chapter. Suffice to say that the following markets, among others, exist in crude oil and refined products:

- crude: the Shell 15-day, Dubai Fatah, Alaska North Slope, Russian Gasoil, Littlebrook Fuel Oil, WTI;
- refined products: Swedish Fuel Oil, Open Spec Naphtha and Boston No 2 Heating Oil.

Each of these markets has its own particular distinctions, customs, and terms, and there are many more local markets.

A classic feature of the oil market is a 'string', 'circle', 'loop' or 'daisy chain' of contracts, under which traders buy and sell the same cargo of oil many times. The sales are back to back, the only difference being the price. This leads to its own logistical problems, which is considered in the chapter on Bills of Lading[11] and in the section on e-trading and chartering, below. Another typical feature of the market is the use of standard forms.

Before considering specific requirements of a trader in a sale contract, it is instructive to focus on a number of legal fundamentals underlying every contract governed by English law.

(a) Evidence of the contract

As a matter of English law, a contract does not need to be evidenced in writing, save for a relatively small number of cases, such as the sale of land. The obvious difficulty with an oral contract is that of proof: a party may, for its own purposes, deny all knowledge of the existence of the contract. For this reason, best practice is for all contracts to be evidenced in writing.

An example of this is a decision of the Commercial Court in 2002.[12] The defendants were the owners of the vessel Epsilon Rosa. They received a recap telex from charterers of the vessel that stated:

---

11 Chapter 8.

owise as perchts standard c/p details amended as per main recap which recap terms to supercede any contradictory terms in the c/p with the full alterations (sic).

The accompanying standard form provided for arbitration in London. In reply, the owners stated in a telex that the recap telex was "in order". The charter of the vessel proceeded and a bill of lading was issued which contained a clause incorporating the terms of the charterparty, including the arbitration clause. A dispute subsequently arose and the owners sought to rely on the arbitration clause. The consignee of the bill (the claimant) sought a declaration from the Commercial Court that the recap telex and accompanying documents did not constitute a duly executed charterparty (and therefore that the arbitration clause was not incorporated into the bill of lading contract). The court held that a charterparty of a vessel could be contained in and evidenced by a recap telex.

(b) Formation

A contract comes into existence when there is agreement between the parties on all the 'essential terms' of the transaction. Generally, the essential terms of an oil trading contract will be:

- the description of the product;
- price;
- quantity; quality; and
- date of delivery.

Mere disagreement or uncertainty as to other general shipping trading terms, such as the laytime/demurrage clause, will not necessarily prevent the contract from being formed.

(c) Consideration

There must be 'consideration': ie, Party A offers to sell one barrel of oil to Party B, and Party B offers to buy that oil for US$1. The consideration in this transaction is the oil and the US$1. In the absence of consideration, there is generally no contract under English law (the exception being contracts executed by deed).

(d) Subjects

If an agreement has been reached between the parties, but 'subjects' remain to be lifted (eg, subject to buyer's board approval), generally no contract exists unless and until those subjects have been lifted.[13]

(e) Implied terms

Terms can be implied into a contract on ground of 'business efficacy or by operation of law. A classic example of terms to be implied in a sale contract are the implied terms in the Sale of Goods Act 1979. These are often excluded in many oil traders' standard general terms and conditions.

12 *Welex AG v Rosa Maritime Ltd* [2002] 1 All ER (Comm) 939.
13 *The 'Junior K'* [1988] 2 Lloyd's Rep 583.

3. Traders' requirements in their trade contracts

3.1 Conflict between terms

Legal disputes over contracts for the sale of crude oil or derivative refined products often arise between parties as to what they have agreed. This is generally for one or both of two reasons.

(a) *Inconsistency between a short form contract and the incorporated main terms with the General or Standard Terms and Conditions ('STCs' or 'GTCs')*

Due to the speed with which trading transactions are concluded, particularly in the 'paper markets' such as Brent Crude and Littlebrook, the use of GTCs has become increasingly common. Use of properly drafted GTCs can have many advantages, specifically that the standardisation of market terms not only avoids a lengthy bargaining process but also enables provision to be made for consequences that the parties may not have had time to address. Conflicts with GTCs can, however, arise where the parties agree specific terms outside of them.

To deal with the problem of inconsistency, the courts have developed rules of construction. The traditional method the English court has employed is the application of a clear rule that, in the event of a conflict between terms incorporated by reference (the GTCs) and the other contract terms, the latter will prevail.[14] It should, however, be noted that this rule only applies where there has been a clear contradiction or conflict between the terms. First principles demand that, where at all possible, the clauses must be harmonised and construed together, unless there is a clear conflict that makes this impossible and a choice must be made. Market practice mirrors legal practice in this respect, as can be seen from a 1991 Court of Appeal case.[15]

Market practice is also developing further to include the use of 'master agreements' in physical trading. In a 'Master Agreement', parties agree to the terms on which they will do business together and then, every time they complete a deal, they enter into a separate transaction confirmation. This development arose originally out of financial trading on ISDA terms. These contracts also frequently contain clauses that have little relevance to the actual physical trade but are fundamental to the commercial relationship, such as the credit clauses.

(b) *Conflicts between terms in the contract*

With regard to this second potential area for dispute, a detailed discussion on English principles of construction of contracts is beyond the scope of this chapter, as is a discussion of the 'parol evidence rule' which is part of that. Traders should nonetheless bear in mind that English law will only permit written and/or oral evidence from negotiations prior to contracting as aids to construction of contracts

---

14 *Sabah Flour Mills Sdn Bhd. Comfez Ltd* [1988] 2 Lloyd's Rep 18 (Court of Appeal).

15 *Phibro Energy AG v Nissho Iwai Corporation and Bomar Oil Inc* [1991] 1 Lloyd's Rep 38. (In a contract for Dubai Crude Oil FOB Fateh, which provided that "where they are not in conflict with the aforegoing, then Conoco general terms and conditions of FOB sales for crude oil will apply." Conoco terms themselves also reflected the prevailing rule of construction, providing "to the extent that any conflict exists between the terms and conditions of these general terms and special terms, the terms and conditions of the special terms shall control.")

in very limited circumstances. Frequently, even these limited circumstances where such evidence may be considered will be expressly excluded if an 'entire agreement' clause has been used in the GTCs.

This does not mean, however, that the courts will construe contracts in a vacuum. As was reaffirmed in a recent decision of the UK Supreme Court,[16] the court's reading of a contract will generally be guided by an overriding principle of 'common sense'. In summary:

> ...
>
> *[T]he ultimate aim of interpreting a provision in a contract, especially a commercial contract, is to determine what the parties meant by the language used, which involves ascertaining what a reasonable person would have understood the parties to have meant. ... [T]he relevant reasonable person is one who has all the background knowledge which would reasonably have been available to the parties in the situation in which they were at the time of the contract. ...*
>
> *It is not ... necessary to conclude that, unless the most natural meaning of the words produces a result so extreme as to suggest that it was unintended, the court must give effect to that meaning. ...*
>
> *[T]he court must have regard to all the relevant surrounding circumstances. If there are two possible constructions, the court is entitled to prefer the construction which is consistent with business common sense and to reject the other. ...*
>
> *Where the parties have used unambiguous language, the court must apply it. ...*
>
> *Since the language of paragraph [3] is capable of two meanings it is appropriate for the court to have regard to considerations of commercial common sense in resolving the question what a reasonable person would have understood the parties to have meant. ...*
>
> *[I]t is not necessary to conclude that a particular construction would produce an absurd or irrational result before having regard to the commercial purpose of the agreement.*

3.2 GTCs

Trades incorporating GTCs need to be on back to back terms. Reference has already been made to the important role of GTCs in crude oil and petroleum products trading contracts. It is often the case that independent traders will adopt the oil majors' GTCs for use by sellers (eg, the BP Terms 2007). It is instructive to look at the application of such terms in certain trading contexts. The following four contexts are instructive of the need to have back to back terms.

(a) Performance

- CFR/CIF ('cost and freight'/'cost, insurance and freight') indicative discharge dates:
  BP Terms 2007, section 7.3:
  Where the Seller expressly or impliedly provides the Buyer with a date or range of dates within which a nominated Vessel shall arrive at the Discharge Port these shall be indicative only, made by the Seller as an honest assessment without guarantee.

---

16 *Rainy Sky SA v Kookmin Bank* [2011] 1 WLR 2900, at paragraphs 14, 20, 21, 23, 40 and 43, per Lord Clarke.

- FOB ('free on board') deliveries:
  Total GTCs 2007, article II.4:
  No stipulation as to time of delivery, whether specified in the Special Terms and Conditions or determined or established in accordance with the procedures set out, or referred to, in the Special Terms and Conditions, shall form part of the description of the Product deliverable hereunder and any obligation to effect or complete shipment by such time is excluded.

Provisions such as article II.4 of the Total GTCs 2007 and section 7.3 of the BP Terms do not pose problems for sellers. They are, however, potentially damaging for those buying on Total or BP Terms (or similar) and selling on different GTCs. It is imperative for this reason that the sale and purchase trades are on back to back terms, as a failure to do so could result in a breach.

(b) Excused/discharged obligations

Force majeure provisions typically provide for contracts to be terminated on account of a prolonged *force majeure* event. These will normally be found in the GTCs. Traders should enquire whether either party can terminate, or whether only the affected party may do so. Further, there are other important questions which traders ought to ask, such as whether the termination is automatic or, if not, after how long termination will take effect.

For example, section 32.4 of the BP Terms 2007 provides for automatic termination:

The appropriate relief under this section shall be as follows:

(i) in respect of an impediment that renders impossible the Relying Party's performance of its obligations, immediate termination of the affected delivery obligation(s) without liability for damages, penalties and other contractual sanctions.

(ii) in respect of an impediment that delays, hinders, reduces or interferes with the performance of the delivery obligation(s), immediate postponement of those obligations without liability for damages, penalties and other contractual sanctions for a period until midnight local time on the last date of the Laydays, or until such time as the impediment is removed, whichever is the earlier…Further, should the impediment continue beyond midnight local time on the last day of the Laydays then it shall be deemed to render the Relying Party's obligations impossible and Section 32.4(i) above shall apply thenceforth…

Automatic termination under this provision creates a potential problem for parties buying on BP Terms and selling on different terms. In such circumstances, there is a real risk of supply contracts on BP Terms being terminated without recourse, with on-sale contracts on different terms remaining enforceable. That is an uncomfortable position for a trader to be caught in. Giving either party of the on-sale contract a right to terminate for prolonged *force majeure* should mitigate automatic termination in a supply contract. The periods after which the right to terminate arises must, however, be back to back.

(c) *Limitation of liability*

There are normally two different approaches to limiting liability: first, fixed cap and, secondly, by reference to an indemnity from a third party.

Clause 63.3 of the LITASCO GTCs 2007 provides a fixed cap:

The monetary liability of LITASCO shall in no circumstances exceed an amount equal to the price of the Oil sold under the Agreement, plus interest (if any due) plus 30%.

Likewise, section XIX.3 of Total's FOB GTCs 2007 provides:

Seller shall in no circumstances be liable for more than the difference between the market price and the contract price with respect to the relevant quantity of Product...

Section IX.4 of Total FOB GTCs 2007 (quantity/quality claims) provides an indemnity cap:

Whenever Buyer is entitled to recover quantity and/or quality costs, losses or damages from Seller, Buyer shall only be entitled to recover such costs, losses or damages from Seller to the extent that Seller is able to recover, and does recover, such costs, losses or damages from Seller's Supplier, and Seller shall not be obliged to pay any amount to Buyer in excess thereof.

Similarly, article 23.A.5 (demurrage) of the Kuwait Petroleum Europoort FOB GTCs 2004 provides:

...Buyer shall not be entitled to recover demurrage from Seller under the Agreement unless Seller is able to recover and does recover such demurrage...from its supplier...and then only to the extent of such recovery.

There is no absolute objection in principle to such provisions, but parties buying on such terms should ensure that they are also incorporated in on-sale contracts. Buyers are often reluctant to agree such provisions, so it is prudent, where possible, to exclude them from the supply contract.

3.3 Recent pitfalls with GTCs

There have been some recent examples of English courts taking a strict approach to GTC drafting. GTCs invariably include a provision akin to section 26.1 of the BP Terms 2007:

...the quality of the Product...shall be not inferior to the specification set out in the Special Provisions...this sub-section constitutes the whole of the seller's obligations with respect to the description, quality and fitness for purpose of the product to be delivered and (save to the extent that exclusion thereof is not permitted or is ineffective by operation of law) all statutory or other conditions or warranties, express or implied, with respect to the description or satisfactory quality of the product or its fitness for any particular purpose or otherwise are hereby excluded. (Emphasis added)

In one case,[17] a seller, Petroplus Marketing, stated in its standard terms:

There are no guarantees, warranties or representations, express or implied, or [sic] merchantability, fitness or suitability of the oil for any particular purpose or otherwise, which extend beyond the description of the oil set forth in this agreement.

Petroplus's GTCs were held not to exclude implied statutory conditions as to product quality under section 14 of the Sale of Goods Act 1979. The court said:

Nowhere in clause 18 is the word "condition" used. [Petroplus] submitted that by

---

17 *Bominflot v Petroplus Marketing* [2009] EWHC 1088.

referring to "guarantees", "warranties", and "representations", express or implied, the clause covered all the different types of contractual terms, including conditions. I disagree. In my judgement, so deeply ingrained is the status of a condition in English law as an obligation the breach of which gives the counterparty the right to terminate the contract or to affirm the contract and sue for damages, that clause 18 is not to be construed as extending to conditions, particularly the conditions implied under s 14 of the 1979 Act.

A 2007 case[18] concerned an FOB sale of gasoline by ERG to Chevron per the ERG FOB GTCs. The Court of Appeal held that the use of the term 'laycan' in an FOB sale contract deprived the buyer of its normal right to terminate and claim damages if the FOB seller failed to deliver within the delivery window. Use of the term 'laycan' allows the FOB seller to cancel the contract, however, if the buyer's vessel does not arrive at the loadport by the end of the laycan.

Generally, there is a further problem with incorporating shipping language into sale of goods contracts. The GTCs (eg, SIETCO's Terms) will make the loading date range a condition of the contract. In a CIF contract, it is for the seller to load within that range. As the seller will charter in the vessel, it will have in its charter a 'laycan' obligation but, in its sale contract, a 'loadport delivery range' obligation. These terms are different animals, but confusion often creeps in and the vessel will pass laycan ETAs to the seller/charterer, who will then pass them to the buyer (often quoting *verbatim* what they have received from the vessel, including the term 'laycan'). The seller/charterer may then request the buyer to agree an extended laycan. When, as is often the case, the vessel arrives outside the loading date range, the market has shifted and it is in the buyer's interest to cancel, the seller will complain, alleging that the buyer's operation staff have varied the contract by agreeing a laycan which, of course (unlike a loading date range), only allows cancellation if the vessel cannot arrive at the port within the laycan spread. Such cases are very fact-sensitive, but highlight how a lack of attention to terminology as well as having terms that are not back to back can cause havoc with trades.

3.4 Description, quality

The description clause, together with the product quantity and quality clause, is a critical one, as a trader will want to know exactly what it is purchasing and, within reason, the exact quantity and specification of the product. A critical issue in any trading contract is to describe accurately the product which the seller is selling and the buyer is buying. As a general proposition, the description of the goods will be interpreted strictly, with the result that even slight discrepancies may be treated as making the goods not what was stipulated. Furthermore, section 13 of the Sale of Goods Act 1979 implies a term into the contract that:

where there is a contract for the sale of goods by description, there is an implied term that the goods will correspond with the description.

It should also be noted that a product can be rejected on the ground that it fails to comply with the description, but is nevertheless of a satisfactory quality.[19]

---

18 *ERG v Chevron ('The Luxmar')* [2007] 2 Lloyd's Rep 542.

19 *MV Bunge v Cie Noga d'Importation et d'Exportation ('The Bow Cedar')* [1980] 2 Lloyd's Rep 601.

This is a critical issue in oil trading, and quality clauses should be drafted carefully to ensure that a trader is selling the correct product. It is quite normal for details concerning the product specification to be incorporated into the trading contract.

Section 14 of the Sale of Goods Act 1979 provides for an implied term that goods "meet a standard that a reasonable person would regard as satisfactory" taking account of all the relevant circumstances, specifically with regard to the goods' fitness for purpose, appearance and finish, freedom from minor defects, safety and durability. Minor deviation from a contract's specifications will generally only give rise to a claim in damages. That said, most GTCs exclude such legislation and their implied terms.

Generally, most crude oil sales are by description, which will often cover the origin of the cargo, eg, "Mellitah Crude Oil of normal export quality as available at original loading terminal at the time or loading ..." or "Sarir Crude Oil of normal export quality etc ..."; (both of which are types of Libyan crude oil).

The sale of products, on the other hand, may not have a description other than the product itself, eg, 'Fuel Oil' or 'Gas Oil'. Sales of product are often on the basis of a quality specification, which lists the parameters to which the product must conform in order to be within specification. Experienced traders will know from the specification, and the quantity, where the product has been refined or from where it originated. However, unless a product says, for example, 'Russian gas oil', then it is not a condition that the gas oil has to originate from Russia. Refineries elsewhere may therefore be able to produce (not necessarily in the same quantity), gas oil having a similar specification to 'Russian gas oil', insofar as that term may have a trade meaning. Where the origin is mentioned, then it is part of the description and only gas oil originating from Russia can comply with the description.

3.5 Quantity

It is usual to find elements of this clause in both the short form contract and the STCs. With regard to quantity, this will usually be a certain tonnage, with a tolerance of a certain percentage.

Section 30 of the Sale of Goods Act 1979 provided that where a seller delivered to the buyer a quantity either less than or more than he contracted to sell, the buyer could reject the goods. The harsh consequences of this rule have been altered by section 4(2) of the Supply and Sale of Goods Act 1994, which states that if the shortfall or excess is "so slight that it would be unreasonable to reject" then the innocent party would only be entitled to damages. Parties can and do deal with these issues in the GTCs, however, and make specific provision as to the quantity delivered being different to the quality contracted for.

The quantity of product loaded onto a vessel will be calibrated by the attending surveyor, who will use the tanker's ullage tables[20] to calculate the quantity which is

20 This is a measurement of space between the surface of liquid in a tank and the top of the tank's inner surface. The reverse of ullage is sounding. This is the depth of liquid in a tank measured from the liquid surface to some reference point at the bottom of the tank. 'Sounding tables' may complement or replace 'ullage tables'.

to be inserted into the bill of lading. The terminal will have its own measuring system which will show the amounts existing in the shore tank(s) before and after loading. There will usually be meters recording the amount passing from the shore line to the ship's receiving manifold. Depending on the terminal or the product, there may or may not be a dedicated loading (and discharging) line.[21] Products loaded from tankers, such as bunkers, will likewise have meters on the exporting and receiving vessels. In some cases, resort is had to the calculation of TPC or TPI (tonnes per centimetre or tonnes per inch) by reference to pre-loading and post-loading draught readings. Every vessel has a stability booklet which in turn includes its hydrostatic curves. From that, the TPC or TPI can be calculated, which informs what total weight of cargo is required to increase the draft by one inch or centimetre.

Disputes and differences can occur on measuring quantities to be inserted into a bill of lading for receipt. First, the volume of liquids is affected by both temperature and pressure. Temperature and pressure volume correction factors for crude oils and refined products need to be applied. This information is available in equations relating volume to pressure which contain physical data that has direct application to volumetric measurement of liquid hydrocarbons.[22] Surveyors sometimes use different tables, which may produce differing results.

The assessment of discharge quantities brings with it another set of problems and propensity for disputes. For certain types of waxy crude and fuel oils, an allowance for ROB ('remaining on board') is permitted. There are in use certain well known oil traders' standard terms for this, such as the 'Sun Oil ROB Clause'. Some products, such as fuel oil, require to be carried at a certain temperature and, in order to do so, the tanks are fitted with heating coils. If heating is not applied, then the convection currents in the cargo slows down or are hindered and the cargo does not circulate within the tank, dropping out into waxy residues which cannot be pumped out. In order to remove non-pumpable residues, lighter products are pumped into the tanks and then the mixture that is liquid pumped out. Crude oil tankers will often use COW (crude oil washing) since the introduction of inert gas systems whereby heated crude oil is sprayed onto the tank walls in order to remove clingage on the tank internal walls. Line hold up and evaporation can also account for differences between loading and discharging figures. The industry recognises the differences between paper and real losses.

In addition there is also VEF (the 'vessel experience factor'). VEF is the historical difference in ship and shore figures for a given ship over a given period, typically based on the previous 10 loadings. It is a factor that indicates the ship's calibration error. The VEF may vary over the life of a ship for a number of reasons. When agreed by interested parties, bill of lading or outturn quantities may be determined, based on vessel received or delivered quantities adjusted by the VEF, which is essentially a correction tool.

---

21 This can be most relevant in connection with off-specification claims.

22 For example Flow-Cal, Inc and the American Petroleum Institute (API) – Chapter 11.1 provides the algorithm and implementation procedure for the correction of temperature and pressure effects on density and volume of liquid hydrocarbons. ASTM (American Standard Test Method) tables are regularly used in the industry.

It is beyond the scope of this chapter to comment further on the technical aspects of loading and discharging measurement of quantities, save to say that it has proved a fertile area for disputes.

3.6 Testing

Given the importance of the quality and quantity issues, it is now standard practice, insofar as it is possible to do so, to test the product at the load port before loading. Various clauses have developed, some of which provide that the test certificate shall be final and binding, while others do not. The seller's requirement will be to insert a clause into the contract stating that the inspection certificates obtained at the load port at the buyer's expense shall be final and binding.

A classic quality and quantity determination clause is as follows:

Certificates of quality and quantity (or load port equivalents) are final and binding on both parties if issued as per standard practice except for fraud and manifest errors.

As a matter of English law, certificates obtained in accordance with the contractual test method are binding.[23] It is, however, important to ensure that the specific contractual test method is used, and no other. The issue of the final and binding nature of quality certificates was considered in a 2002 case[24] in which the contract provided for the analysis to be undertaken in accordance with the ASTM D1298 test method. The quantity and quality determination clause read as follows:

Quantity and quality to be determined by a mutually agreed independent inspector at the loading installation, in the manner customary at such installation. Such determination shall be final and binding for both parties save fraud or manifest errors. Inspector to be appointed by seller. Costs to be shared equally between buyer and seller.

The surveyors instructed used test method ASTM D4052 rather than D1298. D4052 was in fact a more up to date method and it was agreed that even if D1298 had been used the results would have been the same. Notwithstanding this, the buyer argued that the certificate was not final and binding because it was not analysed in accordance with the contractual requirement. Lord Justice Simon Brown agreed, because what was required was a test conducted by the contractual method and none other. The court drew a distinction between a mistake and a departure from instructions. In paragraph 26 of the judgment, Simon Brown LJ said:

A mistake is made when an expert goes wrong in the course of carrying out his instructions. The difference between that and an expert not carrying out his instructions is obvious.

This is a very literal reading of the contract, but is typical of the English courts' approach to the interpretation of contractual terms. It is therefore clear that even though a mistake by an expert does not invalidate the final and binding nature of the analysis and thus the certificate, a departure from what was agreed in the contract does. Hence it is important to ensure that inspection is carried out in accordance with the contractual terms.

---

23 *Toepfer v Continental Grain* [1973] 1 Lloyd's Rep 11.

24 *Veba Oil Supply and Trading GmbH v Petrotrade Inc* [2002] 1 All ER 703.

3.7 Nomination

A further critical issue for traders is the nomination of the vessel under a FOB/CIF contract that can normally be found in the shipping terms dealing with the scheduling of the vessel for loading or discharging. It is important to define precisely the nomination arrangements. Disputes frequently arise out of failure to nominate within time, which will have certain commercial consequences.

A good example of a nomination provision is that in the Conoco General Terms and Conditions, which reads as follows:

> *4.1 Prior to any delivery, the vessel(s) which are to load oil hereunder shall be nominated by the buyer to the seller by written notice (together with the vessel's ETA and the quantity of oil to be loaded)*
>
> *4.2 ...within 5 days of the receipt of the buyer's nomination... seller shall by written notice to buy, accept or reject such nomination.*

This clause was considered in *'The Honam Jade'* case,[25] which concerned a chain of contracts for the sale of 50,000 barrels of Dubai crude oil FOB Fateh for delivery in January 1986. One of the terms of the contract provided:

> *Buyers nomination shall be subject to acceptance by a load port terminal operator and shall conform to all rules and regulations for shipments therefrom.*

Fateh terminal was operated by Conoco at the time and had a limited capacity. The Dubai operator allocated dates for the lifting of parcels of oil to individuals, normally a three-day period except for the last cargo of the month, which was given two days. Nissho made a 'TBN' nomination on December 19 1985 with an ETA of January 15–20 1986. On January 19 1986, Nissho nominated the vessel *'Honam Jade'* with an ETA of January 23–25. On January 23 1986, as Phibro had not responded, Nissho cancelled the contract. The issues before the court were, firstly, the nature of the 'acceptance' which Phibro had to provide and, secondly, the nature of the obligation to accept a nomination within five days – viz whether the term was a condition, an innominate term or a warranty.

As to the first issue, the Court of Appeal held that, on the terms of the contract, Phibro had to give a terminal operator's acceptance, ie, a three-day slot in which the terminal operator had scheduled a parcel to be loaded at the terminal, for the relevant participant at the head of the string, to which the seller had attached the buyer's nomination. In reaching this conclusion, the court was influenced by a number of factors. One factor was that, on its construction of clause 4.1 of the Conoco terms, the acceptance contemplated was not only acceptance of the nominated vessel, but also the proposed ETA and quantity of oil to be lifted. These were matters of great concern to the terminal operator. In addition, clause 4 of the Conoco terms had to be read together with the clause in the contract stating that the buyer's nomination shall be subject to acceptance by a load port terminal operator. Evidence established that this was the standard practice at the Fateh terminal.

With regard to the second question, the Court of Appeal concluded that the term was an innominate term rather than a condition of the contract. As a result, in determining whether there had been a repudiatory breach of contract, the court

---

25 *Phibro Energy AG v Nissho Iwai Corporation and Bomar Inc, 'The Honam Jade'* [1991] 1 Lloyd's Rep 38.

would look at the length of delay and the consequences of the breach. According to the facts of the case, the delay between 14 January (by which time Phibro was obliged to communicate its acceptance or rejection) and 23 January (when Nissho cancelled the contract) was so great that Nissho was clearly entitled to terminate the contract.

Three other cases on the nature of the obligation to give 'notice of a nomination are instructive as they consider standard oil trading terms. The first[26] involved the sale of about 20,000–25,000 metric tonnes of high sulphur fuel oil CIF basis Littlebrook at US$100.75 per metric tonne. In accordance with the standard Littlebrook terms, the contract provided for a "minimum of 3 working days notice of nomination together with the vessel's demurrage rate". On July 15 1987, the sellers nominated the vessel and, under the terms of the contract, the buyers were obliged to open a letter of credit ('L/C') within one working day of the nomination. No L/C was opened and the sellers refused to perform. The buyers complained that the sellers' nomination was non-contractual, arguing that, on its terms, the clause required the sellers to give three days' notice of intention to nominate and that only after these three days had elapsed would the nomination be ineffective.

The Commercial Court held that "sellers to give buyers minimum 3 working days notice of nomination" meant that the sellers were to nominate the vessel three working days before the vessel's ETA at Littlebrook. Mr Justice Phillips said that because there were frequently strings of buyers and sellers acting through intermediaries, and because in 'CIF Littlebrook' contracts the buyer had an option to nominate alternative north European discharge ports, there had to be sufficient time for the notice of the particulars of the carrying vessel and ETA to be passed to the end buyer and for the decision as to the discharging port to be passed back down the chain in due time to allow the course alteration.

The point was considered further in *'The Mathraki'*.[27] In this case Vitol contracted to sell 20,000–25,000 metric tonnes of EEC qualified high sulphur fuel oil complying with Littlebrook specification to Phibro Energy SA "CIF ... basis Littlebrook during the period 1–15 December 1987 at US$103 per metric ton". The case was decided before written standard terms for Littlebrook contracts were published by Amro Oil Ltd. The nomination was given on November 10 1987 at 17:30 London time and was rejected on the basis that the minimum three working days' notice had not been given before the vessel's ETA at Littlebrook. Mr Justice Evans held, however, that three working days meant three calendar working days before the end of the discharge period rather than before the ETA. In disagreeing with the decision in the previous case, Evans J noted that he had the benefit of expert evidence that had been denied to the judge in the earlier case.

The Commercial Court in *'The Mathraki'* also considered the meaning of 'working day' in the Littlebrook market. It was held that a nomination did not have to be made within working hours but could be made up to midnight. The conclusion was reached after the judge heard expert evidence on the market practice which suggested that this was frequently done.

---

26 *Transpetrol Ltd v Transol Olieprodukten Nederland BV* [1989] 1 Lloyd's Rep 309.
27 *Vitol SA v Phibro Energy AG ('The Mathraki')* [1990] 2 Lloyd's Rep 84.

The third case in which the requirements implicit in an obligation to nominate fell to be considered was in *'The Wise'*[28] (which has already been considered above in another context). In this case, Vitol agreed to sell 29,000–30,000 tons of premium motor spirit to Esso at US$22.74 per barrel C&F Hobart and Melbourne. Under clause 12 of the agreement, title and risk were to pass to the vessel's manifold flange at the load port. By clause 14, "Vessel nominated subject to buyer's acceptance. Owners or seller to ensure vessel suitable for discharge at discharge port." The day after the vessel left Bahrain, it was struck by an Exocet missile. At that stage, the buyers had been notified of the vessel's name but they had not accepted it. The buyers, relying upon evidence of market practice, sought to argue that there was a market practice that the passing of title and risk was postponed to the acceptance by the buyer of the seller's nomination. The court held that the custom was contrary to the express terms of the agreement and therefore could not qualify the obligation under the contract. The decision in this case is a clear indication that the court will first of all focus on the actual terms of the contract and not what market practice is alleged to be. It is therefore important to consider the contract first and, if there is some doubt as to its meaning, then to consider what the market practice is.

4. Types of contract form

In international trade, the two contracts that are in general usage throughout the world are the FOB contract and the CIF contract. Classic definitions of these types of contracts are as follows:

> *The FOB contract has many variants but its essential nature is that the seller agrees to place goods which correspond with the terms of the contract on board a ship and to bear all the charges and costs incurred up to and including delivery of the goods on board that ship. The buyer in return agrees to nominate the ship, to pay for the goods, and to pay all the costs incurred after the goods have been loaded on board the ship.*
>
> *A seller under a CIF contract generally assumes more onerous obligations than a seller under an FOB contract in that the seller under a CIF contract agrees to sell the goods at an inclusive price which covers the cost of the goods, their insurance and their freight. The seller is therefore more intimately bound up with the carriage arrangements for the goods than is the case with the seller under an FOB contract.*

There are many other variants on these types of contracts, which Incoterms have made attempts to standardise with a degree of success. Generally, an oil trade will be on FOB, CIF, CFR or DES ('delivered ex ship') terms. However, parties often amend the crucial characteristics of these contracts and sometimes unwittingly transform them into hybrid contracts, as discussed below.

4.1 CIF and hybrid CIF contracts

In *'The Julia'*,[29] Argentinian sellers sold to Belgian buyers 500 tons of rye for shipment 'CIF Antwerp' on the terms of Form 41 of the London Corn Trade Association (LCTA). The agreement provided for "payment to be made against documents".

---

28 *Vitol SA v Esso Australia ('The Wise')* [1989] 1 Lloyds Rep 96.
29 *The Julia* [1949] AC 293.

However, LCTA Form 41 provided:

Condition guaranteed on arrival (subject to any country damaged grain in the fair average quality of the season's crop)...

and

The grain was to be weighed, and seller and buyer were to have the right of supervision over both weighing and delivery, the seller to pay for any deficiency and the buyer for any overweight.

The Umpire of the Board held that it was a "c.i.f. contract" or "c.i.f. business" contract. This was rejected by the House of Lords (as it then was), which held:

...[N]or can a solution be found in the mere designation of the contract as c.i.f.. The true effect of all its terms must be taken into account, though of course the description c.i.f. must not be neglected...

... [W]here the sellers were to pay for deficiency in bill of lading weight, guaranteed condition on arrival and made themselves responsible for all averages, the true view, I think, is that it is not a c.i.f. contract even in a modified form, but a contract to deliver at Antwerp.

So, in that case, although the parties called their contract a 'CIF' contract, it was in essence a DAP ('delivered at place') or DES contract. The result was that the contract was frustrated and the buyer recovered the price.

4.2 CIF and 'Arrival Date' contracts

Three illuminating cases on the nature of such hybrid contracts deserve comment. The first is *'The Wise'*.[30] The terms of the contract provided:

29,000 tonnes to 31,000 tonnes (at sellers' option) of premium motor spirit at Price US$ N C&F Hobart, Tasmania and Holden Dock, Melbourne.

Delivery: Vessel TBN. Arrival March 15–30 1986.

Title and risk: Passes at vessel's manifold flange at loadport.

and

Vessel: vessel nominated to be subject to buyers' acceptance.

The cargo was shipped on board a vessel at Bahrain. The buyer part-paid for the cargo against documents but the vessel was hit by an Exocet rocket while departing the Arabian Gulf. The buyer claimed frustration/breach of condition and the seller cross-claimed for the price. The judge rejected the buyer's claim. The natural construction of the delivery clause (ie, clause 9, "Delivery: Vessel TBN. Arrival March 15–30 1986") was only that the nominated vessel would be expected to arrive in the period mentioned. It was not designed to procure a guarantee of delivery within the period stipulated. On a true construction of the contract, the seller's obligation was held by the court:

...to load a contractual cargo on board a nominated vessel within a time which in the ordinary course of events would enable the cargo to arrive at both discharge ports within the stipulated period, would be accepted.

It is said that so-called c. & f. delivered contracts are a relatively recent development in the oil trade which has come about because of the importance of dates of delivery at the discharge port.

---

30 *Vitol SA v Esso Australia ('The Wise')* [1989] 1 Lloyd's Rep 96.

It seems to me that since the parties have chosen to designate the contract as a c. & f. contract, I should approach the construction of it in the expectation that that is what it is. When I look at the delivery clause, I find that it is to be performed by a vessel to be nominated followed by the words "Arrival: March 15–30, 1986.It seems to me that the natural construction of the clause is that the nominated vessel shall be such as is expected to arrive within the period mentioned. It does not wear the air of a contract designed to procure the guarantee of delivery within the period stipulated. I decline so to construe it, especially since to do so would be inconsistent with a c. & f. contract in classic form.

It is clear from Leggatt J's comments that the label the parties attach to their trade is of importance and indeed the starting point on issues of construction.

The second case is *'The Ballenita'*.[31] The terms provided:

25,000–30,000 tonnes of EEC qualified gasoil meeting Italian winter guaranteed specifications, CIF Genoa.

and

Delivery: on EEC basis in one lot as full or part cargo 1 safe port 1 safe berth CIF basis Genova by M/T "TBN" or sub during October 11–25, 1990. Vessel nomination to be given latest 3 London working days prior to vessel arrival basis Genova.

Judge Diamond QC stated (*obiter*) that, based on "common ground":

Clause 6 in my view provides what characteristics a vessel (whether the vessel "TBN" or sub) must have at the time of nomination. The clause requires that the vessel be one which at the time of nomination can in the normal course of events be expected to give "delivery" between Oct. 11 and 25, 1990.

The third of these cases is *'The Jambur'*.[32] The terms provided:

Sale of gasoil, CIF Kaohsiung

DELIVERY: Latest by 30th April 1990, CIF Basis one safe berth / port Kaohsiung, Taiwan, as full cargo per MT JAMBUR which sailed from Constanza 1000 am 28th March 1990.

It follows that the cargo was purchased afloat. The carrying vessel suffered a collision in the Bosphorus and was arrested after an oil spill. As a result, the vessel could not reach Kaohsiung by April 30 1990. The buyer treated the seller as in anticipatory breach. The seller contended that the goods were at the buyer's risk. The court held that:

the clear meaning of the delivery clause was that the sellers had to procure physical delivery of the cargo by 30 April [emphasis on latest]

In the present case, the delivery clause was a term expressly agreed and incorporated into the contract by the parties. It could only be legitimate to disregard such a clause if it was impossible to give sensible effect to it having regard to conflicting provisions in the remainder of the contract.

It was important to bear in mind that established types of contract for the sale of goods were the servants of those who used them, not vice versa. New trades had grown up and contractual requirements of buyers and sellers of goods could not always be accommodated with the framework of the classic c.i.f. or f.o.b contract.

---

31 *ERG Petroli SpA v Vitol SA ('The Ballenita')* [1992] 2 Lloyd's Rep 455.
32 *CEP Interagra v Select Energy ('The Jambur')* (1990) 289 LMLN 2.

I can see no reason in principle why [parties] should not have agreed that the risk of an untoward delay in the voyage should be borne by the Sellers rather than by the Buyers.

4.3 Conclusions on CIF and 'Arrival Date' contracts

- The mere presence of an arrival date or period on its own will not usually convert a CIF or CFR contract into an 'Arrival Date' contract.
- The court will lean in favour of construing it as a requirement of the vessel to be nominated, *a fortiori* if linked with vessel or vessel nomination – 'label' will usually override.
- The presumption is not, however, hard to displace.
- Any language which emphasises the deadline nature of the date or period will usually be given effect, as will any form of termination right.
- The arrival date or period may also be given a different effect in a contract which is by its terms for the sale of goods already shipped and afloat, for which a *Wise* or *Ballenita* interpretation, looking at the vessel TBN, is less realistic in any event. This was certainly an important part of the court's reasoning in *The Jambur.*

5. Boilerplate clauses

What is 'boilerplate' and why is it important? There is no formal definition, but boilerplate clauses are often present, whatever the subject of the contract. Common examples include:

- governing law;
- confidentiality; and
- *force majeure.*

They are often clauses which deal with the way in which the contract itself operates, rather than the substantive obligations of the parties.

The pitfalls in the usage of such clauses occur because they are presumed to be generic and their importance is often underestimated. Indeed, 'boilerplate' is often used as a pejorative term, but the reality is that such clauses can in fact have significant practical implications for the parties. In litigation, it is often the so-called 'boilerplate' clauses that are frequently overlooked at the time of drafting that end up deciding the result.

It is important to know the background law that lies behind each boilerplate clause. It is a misconception that a lot of things are treated as 'boilerplate' that should not be – for example, penalties/liquidated damages clauses and dispute resolution provisions.

Examples of what could be called an essential checklist of boilerplate clauses are the following:

- 'Entire agreement' clause;
- Representations and warranties clause;
- 'No waiver' clause;
- *Force majeure* clause;

- Confidentiality clause;
- Anti-corruption/compliance with laws clause;
- Penalties/liquidated damages clause;
- Dispute resolution/arbitration clause; and
- Sanctions clause.

5.1 'Entire agreement' clause

A classic example of an 'entire agreement' clause would be framed as follows:

This agreement constitutes the whole agreement between the parties and supersedes all previous agreements between the parties relating to its subject matter…

It is obviously very important and useful where a lot of agreements are made over the course of several telephone conversations, emails or, as is becoming increasingly common, Whatsapp, Yahoo! and other similar instant electronic means of communication.

5.2 Representations and warranties clause

An example of such a clause is as follows:

Each party acknowledges that, in entering into this agreement, it has not relied on, and shall have no right or remedy in respect of, any statement, representation, assurance or warranty…other than as expressly set out in this agreement.

5.3 'No waiver' clause

Such a clause may provide that a waiver must be in writing, and not in respect to any subsequent default. This clause reserves rights in the event of failure or delay in responding.

5.4 *Force majeure* clause

There are a variety of *force majeure* provisions; some set out a long list of *force majeure* events, others are less specific. The latter course carries risks, such as when a credit crunch occurs. A detailed discussion of *force majeure* clauses is beyond the scope of this chapter. Most international oil majors have very similar terms. The term '*force majeure*' has no meaning under English law, so that, on its own, it would probably be void for uncertainty. Most *force majeure* clauses set out a detailed list of events that would excuse liability or suspend performance of a specific obligation under the contract.

There is a common misconception that *force majeure* events must be unexpected or incapable of being anticipated. No doubt this has something to do with the French Law concept of Act of God, from which the term '*force majeure*' was originally derived. Parties often express the view that an anticipated event which may happen through no fault of either is a *force majeure* event and is therefore not something that can result in liability for them. Examples are extreme bad weather, earthquakes, war, terrorism, strikes, breakdown of equipment, fires, explosions, and importantly, inability to obtain the subject-matter of the contract due to lack of supplies, prohibitions by government authorities and suchlike. However, *force majeure* should never permit a self-induced event to justify the suspension of a party's obligations.

Unlike frustration, which operates as a matter of law and leads to the draconian result that, if proven, the contract comes to an end, *force majeure* is a creature of contract and provides a useful way in which to work out a disrupting event without bringing the whole contract to an end. This is, of course, of particular relevance in longer-term contracts that are prevalent in both the upstream and downstream oil and gas industry.

5.5 Confidentiality clause

These clauses are often fairly straightforward, but exceptions will have to be expressly provided for (eg, for the purposes of financing the obligations).

5.6 Anti-corruption/compliance with laws clause

This relates to national laws (particularly those of the US) under trade controls and boycotts. Reputational and corporate governance protection are also relevant considerations.

5.7 Penalties/liquidated damages clause

Such a clause is often treated as boilerplate, but this negates its purpose. The standard formula of percentage of contract in respect of each period of delay may be incompatible with a genuine pre-estimate of loss. Such clauses may therefore be unenforceable and a self-denying prophecy. They must be specific to the contract. They are of more relevance in the offshore construction industry and when considering delay, concurrent delays and such issues, but they should never be treated as boilerplate and should always be the subject of careful consideration and bespoke drafting.

5.8 Dispute resolution/arbitration provisions

By definition, such clauses have a fundamental effect on how disputes are resolved. They are often ignored or treated as a 'midnight clause'. No one wants to talk about disputes when they want to celebrate concluding a new deal: contracts are made to be performed, after all. It may have a fundamental effect on how the contract is interpreted. Such clauses should be bespoke, and consideration given to whether arbitration or the courts are the best forum or whether the parties may want to have the right to elect a particular dispute resolution process depending on how matters develop. It would also be careless, particularly in relation to a long-term contract, not to consider mediation as an option, and the 'arbitration/mediation/arbitration' option should be investigated.[33]

Such clauses should be bespoke. Consideration should also be given to how any judgment or award is going to be enforced. For example, with Chinese-based counterparties, an institutional arbitration award is preferable to a court judgment.

Finally, with the increasing use of multi-tiered arbitration clauses and recent decisions in the English and Singapore High Courts on the enforceability of such

---

33 See, for example, the creation of the Singapore International Mediation Centre (SIMC) and the Arb-Med-Arb clause suggested by the Singapore International Arbitration Centre (SIAC) and the SIMC.

clauses as a condition precedent, great care should be taken when agreeing to 'friendly discussions' or 'attempts to settle' and consideration given to whether they are enforceable obligations to be performed before any arbitration can be commenced.

5.9 Sanctions clause

Such clauses are designed to ensure that appropriate controls and safeguards are in place which are designed to prevent any proceeds or parts of transactions being connected with or finding their way to any restricted person, ie, one who is subject to sanctions in accordance with any trade, economic or financial sanctions laws, regulations, embargoes or restrictive measures administered, enacted or enforced by certain sanctions authorities, such as the Security Council of the United Nations.

6. Traders' charterparty requirements

6.1 Introduction

Generally speaking, traders have developed their own terms on the basis of their experience of particular products and particular load and discharge ports. When chartering a vessel, it is normal to use a standard form (such as the Asbatankvoy) and then add in additional clauses. Traders' standard charter terms are (as would be expected) usually very beneficial to the charterer. Experience has shown, however, that owners frequently sign up to the incorporation of the terms without considering their impact. Generally, the shipping terms in charterparties (such as nomination and laycan) are in the standard form. The focus will therefore be on additional terms that a trader may consider necessary to include/require in the agreement with owners. These include, for example:

- ISM (International Management Safety Code) compliance;
- ITOPF (International Tanker Owners Pollution Federation) membership clause;
- vessel eligibility clause;
- oil major approvals;
- law and arbitration clause;
- cargo retention clause;
- cleaning, cargo degradation;
- blending clause;
- sampling/inspection clause;
- speed and consumption clause;
- laytime and demurrage;
- pumping clause;
- US OPA 90 + US Coast Guard clause;
- STS transfer clause;
- crude oil washing clause; and
- maintenance of cargo temperature clause.

6.2 Standard forms and tanker charterparties of the oil majors

Generally, the contract for the carriage of a product can take one of three forms:

- a voyage charterparty;

- a time charterparty; or
- a contact of affreightment.

Which particular contract is used will depend upon the particular circumstances of the individual trade and trader and the programme the trader is running. A trader with a large portfolio of vessels at its disposal (eg, on long-term charters), may be able to nominate one of those vessels to load the cargo. Alternatively, a trader may have to spot charter a vessel quickly in order to service a requirement to move cargo.

For voyage charterparties, the most frequently used form is the Asbatankvoy 1977. This form, comprising a Part 1 and a Part 2, has been used in the trade for many years. It is a general practice for owners/charterers to add a considerable number of additional clauses which are then incorporated into the charterparty. Over a period of time, owners/charterers have developed from experience a set of their own particular terms and conditions which they seek to incorporate into charterparties; these include, for example, clauses dealing with the ISM Code, oil major approvals, eligibility to load and discharge at relevant terminals and other such terms. For this reason, parties are reluctant to use new forms to which they are not accustomed. Old forms therefore remain popular and new forms take some time to obtain significant market usage (and often do not). However, other forms do exist, such as BP Voy 4. This document was developed by BP in conjunction with Intertanko. It is 40% longer than BP Voy 3 and has sought to incorporate various terms and conditions that BP normally likes to use. For this reason, it is generally considered to be weighted in favour of charterers. Other forms also include Asba II, Tankervoy, Shellvoy and Mobilvoy. The most popular voyage charters are the Asbatankvoy and the BPVoy.

With regard to time charterparties, the most popular time charter forms in the oil trade are the Shelltime and BP time forms.

7. Conclusion

The common law tradition is closely tied to crude oil and refined product transacting, with English and US laws historically the preferred choice of law for many in the industry. This trend is likely to continue, as countries like the US continue to account for an increasingly substantial proportion of worldwide production and consumption.

The specific challenges involved in transacting under English law, as discussed above, show how just important it is to ensure that the trading contract appropriately reflects the parties' intentions. This begins with the form of contract chosen and extends to the wording of the clauses, such as a quantity and quality determination clause. Furthermore, it is essential not to underestimate the importance of boilerplate clauses, as they will often play a vital role in deciding which way a dispute is resolved.

Delivery and nominations

Stuart Shepherd
Ince & Co LLP

---

1. Introduction

This chapter discusses contractual obligations with regard to delivery under oil and gas contracts, together with notices of nomination, which are often called for under such contracts. In this context the word 'delivery' means physical delivery of the goods concerned This chapter also comments, however, on the tender of documents under oil and gas contracts since 'delivery' in the wider sense would, under certain contracts, not be complete until there has been both physical delivery of the goods and tender of documents relating to the goods.

Almost invariably, oil and gas contracts will stipulate a period within which the goods must be delivered, whether by way of shipment on board the carrying vessel at the loadport, discharge from the carrying vessel at the discharge port, or otherwise. In most contracts, the respective obligations on the seller and buyer to comply with such requirements with regard to the timing of delivery are conditions of the contract, breach of which will permit the other party to terminate the contract. Apart from considering these contractual obligations, this chapter also discusses the obligations of both the buyer and the seller to give notices of nomination connected with the delivery of the goods, including nominations required from an FOB ('free on board') buyer in respect of the vessel nominated to load the goods, and nominations required from CFR/CIF/DAP ('cost and freight'/'cost, insurance and freight'/'delivered at place') sellers in relation to delivery by shipment on board the carrying vessel at the loadport or delivery at the discharge port.

2. Delivery

2.1 Delivery under different types of contract

(a) FOB contracts

Delivery under an FOB contract requires collaboration between the buyer and the seller. It is the buyer's obligation to procure a vessel to load the goods at the loadport and the presentation of the vessel there to load triggers the obligation on the seller to put the goods on board the vessel nominated by the buyer. The seller is not obliged to have the goods ready to load until the buyer's nominated vessel is ready.

In terms of timing, under a 'classic' FOB contract the buyer is obliged to nominate a vessel to present at the loadport capable of completing loading within the delivery period – this being the period within which the seller is obliged to put the goods on board the vessel. This will inevitably require the vessel to be ready to load at the loadport some time before the end of the delivery period. How long in advance of the end of the delivery period the vessel will need to arrive will be equal to the period reasonably required to load the cargo on the vessel concerned. Assuming that the buyer complies with these obligations, the seller's obligation under a 'classic' FOB contract is to put the goods on board the vessel within the delivery period. Since the time for taking and giving delivery is of the essence of an FOB contract, any breach of the buyer's or seller's respective obligations would expose the defaulting party to a claim for damages from its counterparty and would entitle it, if it so wished, to refuse to take or give delivery, as the case may be.

The delivery position under a 'classic' FOB contract may be, and often is, modified by express terms in the contract. For example, the contract may merely require the buyer's nominated vessel to arrive at the loadport ready to load at any time within the delivery period. Further, parties often agree a loading 'laycan', which is not the same as a delivery period. The consequences of the vessel not arriving within a laycan, and of the inability of a seller to put the goods on board the vessel within such period, differ by comparison with the position under a delivery period. The term 'laycan' derives from charterparties. It is a period of time in which a vessel is expected to arrive at the loadport. Under a charterparty, if the vessel arrives before the beginning of the laycan, laytime will not start to run until the beginning of the laycan. Further, if the vessel arrives after the end of the laycan, the charterer will have the option of cancelling the charterparty. In the context of a sale contract, if an FOB buyer's nominated vessel arrives before the beginning of the laycan, the laytime allowance under the sale contract will generally not begin to run until the beginning of the laycan. More importantly, if the vessel arrives after the end of the laycan, the seller will usually be entitled to cancel the shipment. Further, unlike the position under a delivery period, should the buyer's vessel arrive before the end of the laycan (and if required by the contract in sufficient time to load the cargo before the end of the laycan), but the seller is unable to deliver the cargo, the buyer will not be entitled to terminate the shipment. He will be limited to a claim to demurrage for the period of delay in loading once the laytime period has expired.

The use of laycans in this context raises the interesting question as to what is the seller's obligation to ensure that the nominated vessel arrives at the loadport within the laycan. Can he simply decide not to bother trying to put a vessel in with impunity, knowing that the worst that can happen is that the seller will be entitled to cancel the shipment? While this remains a moot point, in most charterparties there will be an implied term requiring the vessel owner to commence the approach voyage to the loadport at such time as it could reasonably be anticipated that the vessel would arrive within the laycan. It is arguable that a similar term should be implied into a sale contract where the timing of delivery is expressed in terms of a laycan.

In one sense, delivery of the goods on board the vessel provided by the buyer does not deliver the goods to the buyer, since the party who will be entitled to delivery of the goods at their destination will be the holder of the bill of lading, as (absent any special arrangement between the buyer and the carrier, such as the provision to the carrier of a letter of indemnity against delivery without production of bills of lading) only the holder of the bill of lading will be entitled to take delivery of the goods at the discharge port.

In terms of tender of documents under an FOB contract, different species of such contracts have been identified; they are distinguished largely by the nature of the obligation on the seller to procure a bill of lading and on what terms. Under some FOB contracts, the seller has no responsibility to procure a bill of lading. Under others, the seller does have such a duty. In order to determine the species of the FOB contract concerned, however, it is necessary to identify whether the bill of lading that must be procured and tendered has to be issued in the buyer's name and to the buyer's order or may be issued to the seller's name or order. Where the contract provides for payment against presentation of (among other things) a bill of lading, the inference would generally be that the contract is one which falls in the latter category. In most cases there would be an obligation on the FOB seller to obtain a bill of lading and thus to tender it, together with any other documents required by the contract, to the buyer. In those circumstances, the obligation upon the seller to tender documents will be similar, although not identical, to the obligations upon the seller under a CFR/CIF contract; these are discussed in more detail below.

(b) CFR/CIF contracts

Under both CFR ('carriage and freight') and CIF ('carriage insurance and freight') contracts, it is the seller who arranges the carriage of the goods to the discharge port. Further, under a 'classic' CFR/CIF contract, the seller must load the goods on board the vessel within any contractual delivery period. He must also procure proper shipping documents, including bills of lading, and tender them to the buyer or as stipulated by the contract (for example, by tender under any letter of credit). Importantly, the seller is not, however, under any obligation to ensure that the goods reach their destination. Once the seller has shipped the goods on board the vessel and tendered appropriate documents, his delivery obligations are at an end.

Most CFR/CIF oil and gas trading contracts, however, contain delivery provisions focused on the arrival of the goods at their destination and contain no express reference to a period during which the goods must be loaded on board the vessel at the loadport. The wording of such contracts does not always, and in fact quite often does not, impose any obligation on the seller to ensure delivery at the discharge port within the dates stated in the contract, as very clear words are required to 'convert' the seller's obligation under a CFR/CIF contract from that of shipping the goods to actually guaranteeing their arrival at the discharge port. For example, in *P&O Trading Ltd v Scanoil Ltd*,[1] a CIF contract provided for "delivery February 15/March 15 basis Rotterdam". The court held that this wording did not impose an obligation on the seller to ensure delivery at the discharge port within that date range – or indeed at all. It held that the provision merely meant that the seller had to load the vessel in time to enable it to discharge at the discharge port within the specified period, but was not responsible if delivery at the discharge port were delayed beyond that period by virtue of circumstances outside the seller's control.

---

1 [1985] Lloyds' Rep 389.

A contract in which the wording is sufficiently clear to amount to a guarantee by the seller that the carrying vessel will arrive at the discharge port at such time as to permit complete delivery within the delivery period – or simply to arrive by a particular date – is usually referred to as a 'CIF delivered' contract.

As with FOB contracts, parties to CFR/CIF contracts sometimes use the term 'laycan' when referring to the anticipated loading period. In that context, if the nominated vessel fails to arrive at the loadport within the contractual laycan, the buyer would be entitled to reject the shipment. Whether he would also have a claim for damages would depend on the reason for the failure to meet the laycan and whether, in the circumstances, it was possible to imply some form of obligation on the seller to make the necessary arrangements to meet that laycan.

As can be seen from the above, a CIF/CFR seller's obligations in respect of delivery can, depending on the precise wording of the contract, range from:

- merely loading the goods on board the carrying vessel at the loadport within a particular date range;
- arranging loading at the loadport at such time that, all things being equal, the vessel will be able to arrive at the discharge port within a given period; and
- guaranteeing the delivery of the cargo at the discharge port within a specified period.

Traders often deploy other, often relatively opaque, ways of expressing the seller's delivery obligation. Examples of these are contracts in which the timing of delivery is defined by reference to scheduled loading dates or require discharge 'consistent with' specified loading dates. Such provisions come in such diverse forms that general guidance as to their construction is difficult. The courts will, however, imply obligations of good faith where necessary and appropriate. For example, in relation to a provision referring to delivery by reference to specified scheduled loading dates, the court may find that there is an implied term requiring the seller to have an honest belief that the particular cargo was scheduled for loading within the dates concerned.

(c) DAP contracts

DAP ('delivery at place') contracts usually impose on the seller an obligation to ensure delivery of the cargo at a particular place within a specified delivery period. Alternatively, if so worded, they may require the seller's nominated vessel to arrive at the discharge place within such period. The Incoterms 2010 provide, in relation to DAP contracts, that the seller must deliver the goods by placing them at the disposal of the buyer ready for unloading at the agreed place within the agreed period. Therefore, if the contract is based on the Incoterms, the seller satisfies his delivery obligations when the delivering vessel tenders a valid NOR at the discharge port/place within the delivery period.

2.2 Interaction of delivery terms with other contract terms

(a) *Letters of credit and payment security*

It is important to apply contract delivery provisions in the context of the sale contract as a whole and not in a vacuum. The most obvious illustration of this is in the context of contractual provisions in relation to payment security, in particular, the provision of a letter of credit. Absent contrary provisions in the sale contract, the seller's obligation to deliver the goods will be conditional upon the provision of such payment security. To illustrate the point, the seller under an FOB contract would not, absent contrary provisions, be obliged to load the buyer's nominated vessel until any letter of credit or other payment of security called for under the contract is in place.

Ideally, the contract will specify the date by which the payment security must be in place. It may specify a particular date or a period before the first day of the delivery period by which the payment security must be provided. If it is not possible to derive such a deadline from the contract, a term will be implied as to when such payment security should be provided, depending upon the nature of the contract and its terms. Almost invariably in such circumstances, that deadline will be at the very latest immediately before the commencement of the delivery period and sometimes a reasonable period before then – the philosophy being that the buyer should be entitled to a reasonable period of time following receipt of the contractual payment security to make the goods ready for shipment.

As with the obligations with regard to delivery, the obligation to provide payment security by a particular time is also, almost invariably, a condition of the contract, such that if it is not provided on time, the seller can treat the buyer's breach as repudiatory and terminate the shipment. This is something that is not always appreciated and thus overlooked. It is important that such obligations are not overlooked because, while it remains the position that, until the payment security is in place, the seller has no obligation to deliver, the seller can lose that right to terminate the contract. If the seller does or says anything in such circumstances which communicates to the buyer that he will not terminate the contract for such breach, the seller may waive the right to terminate and will remain obliged to deliver the goods once the payment security is put in place (except where, for example, the delay is extreme and the contract is frustrated).

(b) Force majeure

This chapter is not the place for a detailed discourse on *force majeure*. If the absence of payment security will not 'excuse' a failure to deliver goods within the relevant delivery period, it is possible that there will be a *force majeure* clause in the contract that will. *Force majeure* clauses come in different shapes and sizes; some are generous, others less so. While they are exception clauses, they are, like any other contractual provision, to be given their plain and ordinary meaning. If, in the context of delivery, a seller can establish that the factual situation preventing delivery falls with the definition of '*force majeure*' in the contract, the clause will protect the seller so long as he complies with any notice requirements contained in it. More often than not, however, sellers find that *force majeure* clauses do not give them the protection they anticipated or hoped for. One important factor is that it is necessary to show that a *force majeure* situation has indeed arisen such as to prevent the seller from performing the contract and that, in general terms, but for a *force majeure* event, he would have been able to perform it. It is also necessary to show that the *force majeure* circumstances did not result from the seller's own breach. By way of example, some generously worded *force majeure* clauses provide protection in the event of a vessel's breakdown on the voyage or even delays in the voyage. Leaving aside the potential difficulties of establishing 'delay' in that context, a *force majeure* clause in such terms will not protect the seller under a CIF delivered contract if, say, the vessel, while in fact delayed by bad weather, would never have been able to give delivery at the discharge port within the delivery period in any event at the time the vessel departed from the loading port.

3. Nominations

3.1 Introduction

Oil and gas sale contracts generally contain nomination provisions. Such provisions can relate to the cargo to be delivered under the contract and/or the vessel that is to take or give delivery, depending on the type of contract concerned. Contracts, in particular term contracts, usually have different clauses dealing with the nomination of the cargo and of the vessel that is to load/deliver the cargo. Term contracts, in particular, often have provisions requiring an exchange of formal notices with regard to the quantity and timing of each delivery. For example, a contract that calls for one cargo delivery per quarter may provide for the seller to nominate the month of delivery some time in advance of that month, and then send a further nomination at a later date narrowing the delivery period within that month. Equally, it may be for the buyer to respond with his own nomination narrowing the delivery period. Quite separately, there may be a provision dealing with the nomination of the performing vessel that requires the nominating party to provide certain information in relation to that vessel and to seek the counterparty's approval of it. So far as nomination provisions in relation to the goods themselves are concerned, these are generally designed to give the buyer notice of the shipment of the goods and, in relation to CIF delivered and DAP contracts, to narrow the delivery period so that the buyer knows when to expect delivery of the goods, so that the necessary arrangements can be made to take delivery of them. On the other hand provisions focused on the nomination of the performing vessel are designed to ensure that the vessel is one that will be able to deliver the goods at the discharge port.

Most oil and gas sale contracts are for the sale and purchase of unspecified goods meeting a particular description, the obvious exception being where the contract refers to existing goods on a particular vessel. In the usual case of a sale of unspecified goods, the goods will at some point be identified and appropriated to the contract. At the time of appropriation, the contract ceases to be one for the delivery of unidentified goods and becomes a contract for delivery of specific, identified goods. Specific goods can be appropriated to the contract in different ways, but one way in which this is done is through the nomination procedure provided for in the contract or by way of acceptance of a tender of documents. Generally, once the goods have been appropriated to the contract, they may not be unappropriated unless the contract provides a mechanism for such 'unappropriation'. As discussed below, however, some contracts do allow for this through express rights of substitution.

3.2 Nominations and the tender of documents under the different types of contract

(a) FOB contracts

Unless there is contrary provision in an FOB contract, the buyer has the right to nominate an appropriate vessel at any time, subject to there being sufficient time from the nomination to load the cargo by the end of the delivery period. If the buyer fails to give a nomination in time, the seller would be entitled to refuse to deliver the cargo and to claim damages. FOB contracts will, however, usually expressly provide for a period of notice prior to loading that the buyer is required to give to the seller. Any failure by the buyer to give such notice period before the end of the contractual delivery period will entitle the seller to refuse to give delivery. In *Bunge Corporation v Tradax Export SA*,[2] the contract concerned US soya bean meal and called for shipment in 5,000 mt lots in May, June and July 1975 FOB US Gulf port at seller's option. The contract required the buyer to give the seller 15 days' notice of loading. The May shipment period was extended so that, in effect, that shipment became the June shipment. Notice of loading was given by the buyer on June 17 1975 and the seller said that it had, therefore, received only 13 and not 15 days' notice before the end of the shipment period and was thus entitled to refuse to deliver that shipment. Quite plainly, the seller was right to say that it had been given only 13 and not 15 days' notice before the end of the shipment period, but the question was: what was the contractual consequence of this? The case eventually reached the House of Lords, which had to decide whether the requirement to give 15 days' notice of loading was a condition of the contract, breach of which excused the seller from performing the contract, or a mere warranty, breach of which only gave the seller the right to any damages caused by the 'short' notice. The House of Lords held that the requirement was a condition, so that the seller was entitled to terminate the shipment.

The nomination provision in the contract may also specify certain information that must be included in the nomination. This is often expressed in mandatory terms, eg, "such notice shall specify ...". It is suggested that if the notice does not include the specified information, it will be invalid.

(b) CFR/CIF/DAP contracts

Under a CFR/CIF contract, a seller's primary obligation, having shipped or bought goods afloat, is to tender to the buyer the shipping documents required by the contract. If the contract is silent as to the documents to be tendered, the seller is bound to tender three documents, namely:

- a bill of lading;
- a policy of insurance (in the case of a CIF contract); and
- an invoice.

---

2 [1981] Lloyd's Rep 1.

Oil and gas sale contracts, however, normally call for the presentation of various additional documents, such as certificates of quality, quantity, origin and the like. The seller will not have complied with his delivery obligations under the contract until the documents required are tendered.

There is a good deal of law dealing with the requirements which such documents must satisfy, in particular in relation to bills of lading. This chapter does not intend to examine this area in any great detail. So far as the bill of lading is concerned, however, a fundamental principle is that the bill should provide what is called continuous documentary cover. In other words, the bill of lading, upon transfer to the buyer, must be in a form which entitles the buyer to delivery of the cargo at the contract destination. Thus, for example, if the seller intended to satisfy his delivery obligations under the contract by shipping goods from one port, then transhipping them elsewhere – say, at an offshore location – prior to a second vessel loading them for delivery at the CIF destination, the bill of lading issued at the first loadport (for delivery at the transhipment destination) would not provide continuous documentary cover. This is because possession of such bill would not entitle the buyer to delivery of the goods from the carrier at the CIF destination.

Contracts will normally contain express stipulations as to the time at which documents must be tendered. Where, however, the contract contains no such provisions, the seller must tender the documents as soon as possible after the goods have been shipped, or, in the case of goods bought afloat, the seller has allocated the goods to the contract.

Contracts which contain stipulations as to the time of tender of documents can fall into two categories: those which specify a precise time for tender and those which do not. An example of the former would be a provision requiring tender of documents "no later than seven days after the bill of lading date", while one of the latter would be a stipulation requiring tender "without delay". While there are differing views, and each case will turn on the particular wording of the contract, it is likely that a provision calling for tender of the shipping documents by a precise time will be treated as a contractual condition entitling the buyer to reject a tender that is not made in time. By contrast, a provision which contains imprecise terms with regard to the timing of tender may amount to an innominate term, meaning that a slight delay in tendering the documents would not permit rejection, whereas a lengthy one might. There appears to be no clear legal authority on the point. The position, it is suggested, is the same where the contract provides no express provisions regarding the timing of tender where, as indicated above, it is implied that the tender will be made as soon as possible after the goods have been shipped or, in the case of goods bought afloat, allocated by the seller to the contract.

In addition to these obligations with regard to the tender of documents, oil and gas contracts normally contain a provision requiring the seller to "declare" a shipment made under the contract by giving a notice of nomination. As mentioned above, such notice is designed to give the buyer early notice of the shipment that has been appropriated to the contract, so as to facilitate resale and to enable the buyer to make the necessary arrangements for discharge of the cargo at the discharge port.

The notice of nomination must be given within the time stipulated in the contract and, generally, such requirement will be considered a condition of the contract, such that a failure to give a timely notice of nomination will permit the buyer to reject that nomination. Thus, for example, if the contract required a notice of nomination to be given within seven days of the bill of lading date, a nomination given eight days after that date would be invalid and open to rejection. A buyer needs, however, to be aware that an invalid nomination is a nullity and the seller may be able to make a new, valid nomination in respect of a different shipment. It is therefore important that a buyer does not prematurely terminate a shipment or contract for repudiatory breach upon receipt of an invalid nomination; he should only terminate in circumstances where it is no longer possible for the seller to make a valid nomination.

In *Vitol SA v Phibro Energy AG (The Mathraki)*,[3] the contract required the seller to give "3 working days' notice of nomination". This was in the context of a CIF delivered contract which specified a discharge port delivery period of November 1–15 1987. The combination of these provisions meant that the seller had to give a notice of nomination at least three working days before the end of the delivery period. In the circumstances, the buyer (Phibro) could reject the shipment and, as this was a spot contract, terminate the contract for breach if it had not received a valid nomination three working days before the end of the delivery period. The nomination had been given at 18:30 on the last day upon which a valid nomination could be given and the issue was whether the notice had to be given during 'working hours' or at any time up until midnight. The court concluded that the seller had until midnight, so that the nomination had been made in time. The alternative case run by the buyer was that it was entitled to reject the nomination on the ground that the estimated time of arrival given in the nomination was less than three working days prior to the vessel's arrival at the discharge port. The court also rejected this argument, concluding that the contract terms were structured so that the buyer was to receive notice of the vessel's nomination rather than of its ETA at the discharge port. At most, Phibro may have been able to resist any claim for demurrage that arose due to any delay in discharge between the time of the vessel's arrival at the discharge port and the expiry of the three working days' notice of nomination.

Apart from complying with contractual provisions with regard to the timing of the giving of notice of nomination, the contents of the notice must comply with the contractual requirements. They must contain all the information required by the contract and that information must be accurate. Absent or inaccurate information will render the nomination invalid. If, however, the seller includes in the notice gratuitous information not called for by the contract, the notice will not be invalid even if that further information is inaccurate. It is important for a buyer to voice promptly any objection it wishes to raise in respect of any defect in a notice of nomination, whether the defect is in terms of timing or content. If a buyer acts in a manner consistent with acceptance of a defective notice, it may be taken to have waived the defects in it and so lose the right to reject it.

3 [1990] 2 Lloyd's Rep 84.

(c) ***Vessel approvals***

A buyer's nomination of the loading vessel under an FOB contract will obviously identify the vessel, as will a notice of nomination under a CFR/CIF/DAP contract. Further, shipping documents tendered to a buyer will identify the vessel upon which the goods have been loaded. So long as the vessel complies with any express and implied requirements in the contract, the buyer is obliged to accept the nomination/tender. Many oil and gas sale contracts, however, contain express provisions which either require the buyer or seller (as the case may be) to obtain the approval of the counterparty to the vessel concerned and/or permit the counterparty to reject any vessel nominated by its counterparty. Sometimes, in the case of the CFR/CIF/DAP contracts, the contract is structured in such a way that the seller has to obtain the buyer's approval of the proposed vessel in advance of a subsequent nomination of the shipment of the goods on that vessel. Such clauses often identify the circumstances in which the counterparty can reject a vessel that has been nominated; others simply provide, for example, that acceptance cannot be unreasonably withheld. The latter provision is difficult to apply in practice, as there will be different interpretations of what reasonableness may dictate in any particular situation. Further, such clauses will normally require the party receiving the nomination to raise any objection to the nomination within a given period of time, often 24 hours. This time limit is important since, if the vessel is validly rejected, the nominating party will need to nominate another vessel or shipment. One seldom sees, however, a provision which automatically extends the time for making a further nomination in circumstances where there is a valid rejection of the initial vessel nomination; this is a factor which the nominating party should bear in mind in terms of the timing of its nomination. If there is a risk that a nomination might be validly rejected, the nominating party should seek to ensure that it has sufficient time to make a further, valid nomination in time.

(d) ***ETA notices***

Oil and gas sale contracts sometimes also include, in the case of CFR/CIF/DAP contracts, provisions requiring the seller to give notice to the buyer in respect of the vessel's ETA at the discharge port. Such obligations are not generally regarded as being in the same category as nomination provisions, which are treated as fundamental terms of the contract. In other words, while a failure to give an ETA notice will amount to a breach and thus expose the seller to a claim for any damages suffered as a result of the failure to give the buyer such notice, such failure will not generally entitle the buyer to terminate the shipment or contract concerned.

(e) ***Substitutions***

There are various circumstances in which a buyer or seller (as the case may be) may wish to make a new nomination in place of a previous, valid nomination. Whether this is permissible will depend on the terms of the contract but, generally speaking, seeking to reserve one's right to nominate a substitute by adding words such as 'or sub' in a notice of nomination, will not be effective to give the nominating party a right to substitute unless the contract entitles it to do so.

In the context of an FOB contract which contains no particular terms in relation to the timing and content of the buyer's nomination of the loading vessel, it is debatable whether, once the buyer has nominated the vessel, he is entitled to nominate a substitute.

In *Cargill UK Ltd v Continental UK Ltd*,[4] which concerned an FOB contract, Evans J commented:

> *There is no general rule of law, in my judgment, which entitled the buyers to substitute another vessel in place of one that has been duly nominated and no such right is reserved by the terms of the contract ....*

The contract may, however, contain express nomination provisions which identify whether the buyer can substitute, or which, in the circumstances of the case, rule out the possibility of substituting. The *Cargill* case was an example of the latter. In that case, the buyer had to give a provisional notice of nomination eight clear days before the vessel's ETA which was to be followed by a final, definite notice four clear days of the date of the presentation of the vessel for loading. That date had to be a maximum of seven days from the original ETA given in the provisional notice and had to be before the end of the contractual delivery period. The buyer initially gave provisional notice nominating the "*Cobetas* or Sub" and, two days later, gave a definite notice in respect of the same vessel with a slightly later ETA. Four days later, and five days before the end of the delivery period, the buyer gave a further notice purporting to substitute the "*Finn Beaver* or sub" for the *Cobetas*. Despite the fact that this notice was given four clear days before the end of the delivery period, the court held that the purported substitution was invalid. It went on to hold that both the provisional and the final notice had to be given in respect of the same vessel and, even if there was a general right to substitute (which, as indicated above, the court doubted) it was, on any basis, given the express terms of the nomination clause, too late for the buyer to substitute.

In the context of CFR/CIF/DAP contracts, once a cargo has been nominated it is submitted that that nomination cannot be substituted unless the contract contains express provisions permitting substitution. This is for various reasons, not least the fact that, by nominating specific goods under the contract, the seller is appropriating the goods to the contract and the goods become specific to it. Oil and gas contracts sometimes expressly permit substitution, however. In *ERG Petroli SpA v Vitol SA*,[5] the court had to consider at what point an express contractual right of substitution could be exercised. The contract in that case provided that delivery was to be "one safe berth one safe port c.i.f. Genoa by MT 'TBN' or sub during October 11–25 1990. Vessel nomination to be given latest three London working days' prior to vessel's arrival basis Genoa". On October 7 1990, the seller nominated "*BP Energy*/sub". The *BP Energy* subsequently arrived at Genoa and gave NOR. The cargo on board the *BP Energy* was, however, found to be off-specification and the seller, on October 19 1990, nominated the Ballentina in substitution for the *BP Energy*, but the buyers refused to accept the second nomination. It was common ground that the words "or sub" in

4 [1989] 2 Lloyd's Rep 290.
5 [1992] 2 Lloyd's Rep 455.

the contract conferred on the seller a right to substitute. There was, however, a dispute as to when the option to substitute had to be exercised by the seller, the buyer's complaint being that the substitution was invalid because the substitute nomination was not given three days prior to the vessel's arrival in Genoa. As in *The Mathraki*,[6] the court concluded that, while the failure to give the substitute nomination at least three days before the vessel's arrival in Genoa was a breach, it did not invalidate the substitution. The position would have been different, however, if at the time the substitute nomination was given it would not have been possible to discharge the cargo within the delivery period on the expiry of three working days from receipt of the substitute nomination.

Finally, in the context of CFR/CIF contracts, what is the position if there is no express contractual requirement to nominate the goods? In such a situation, the goods are likely to be appropriated to the contract through the tender of the documents under the contract. Assuming the tender is compliant with the contract and thus valid, it will not be open to the seller to make a substitute tender unless the buyer rejects that tender or agrees to the substitution.

6 Note 3 above.

Carriage

Ben Olbourne
39 Essex Street

---

1. Introduction

Vast quantities of oil and gas products are transported around the world, from source to refinery to end-user with potentially many stops in between. This chapter considers the terms pursuant to which such products are typically carried. It necessarily does so in very broad outline only: given the diversity of the products, many of which require their own specialised means of carriage, and the diversity of terms under which these products are carried, a more detailed analysis would far exceed the purpose and parameters of this chapter.

The focus of this chapter is principally on the carriage of oil and gas products by sea, although some brief comments are also proferred in relation to carriage by other means, namely pipeline and rail/road. The author's intention is to highlight the key legal features associated with the carriage of these cargoes, primarily from a practical and commercial rather than an academic perspective. Certain topics have, however, been omitted or are dealt with only briefly; either they are the subject of other contributions in this book or they are more concerned with matters of general contract law and not, more specifically, with the carriage of oil and gas products. Topics falling into this latter category include: contract formation and interpretation; breach of contract; justifications for non-performance, including illegality and frustration; and remedies, including termination and damages. Other omissions may be justified only by reference again to the purpose and parameters of this chapter. All matters of law are considered from the perspective of English law.

2. Carriage by sea

Generally, the carriage of a cargo by sea is the subject of two different contracts. One is the charterparty (or charter) between the owner of the vessel[1] and the charterer: the other is the contract concluded between the carrier, who may be either the owner or the charterer but is typically the former, and the owner of the cargo being carried. This chapter is concerned only with the first of these types of contract. The latter type, which is routinely contained in or evidenced by a bill of lading, is the subject of another chapter.[2]

Charters may take many forms, the principal ones being time charters and

---

1 The 'owner' for the purposes of a charter may not in fact be the ultimate legal or beneficial owner of the vessel, but may itself be a charterer of the vessel.

2 See chapter 8, *Bills of lading*.

voyage charters, both of which are considered in detail below. Contracts of affreightment are also considered in this chapter as they are often closely associated with voyage charters and, indeed, are typically umbrella contracts covering the performance of a series of voyage charters. Other forms of charter, including bareboat charters, are of lesser significance and are not frequently seen in the oil and gas context.

2.1 Time charters

(a) *Key features*

A time charter is, in essence, a contract whereby the owner of a vessel transfers to the charterer "the right to exploit the earning capacity of the vessel".[3] The owner retains possession of the vessel and continues to operate it but, for a defined or determinable period of time and in exchange for the payment of hire, agrees to comply with such directions in relation to the use of the vessel as the charterer may give consistently with the terms of the charter.

In general, although in every case necessarily subject to detailed consideration of the precise terms of the charter in question, the parties' respective principal obligations under a time charter are as follows.[4] The owner generally undertakes:

- that the description of the vessel given in the charter is accurate at the time the charter is concluded;
- to 'deliver' the vessel to the charterer (ie, by making it available for performance of the charterer's orders) by the date stipulated in the charter and in a seaworthy condition;
- that, thereafter, it will comply with the charterer's orders as to the employment of the vessel, provided that those orders are consistent with, among other things, the provisions of the charter governing the vessel's geographical trading limits and the range of permissible cargoes;
- to cooperate in respect of the loading and unloading of cargoes;
- in order to facilitate the charterer's commercial use of the vessel, to enter into contracts with third parties for the carriage of goods loaded on to the vessel (typically bills of lading) on such terms as the charterers may require;
- throughout the charter period, to meet the vessel's running expenses save in relation to fuel (ie, bunkers); and
- to carry out necessary cleaning of the vessel between cargoes and to maintain it in a fit condition for service.

On the other hand, a time charterer generally undertakes:

- to pay hire for the duration of the charter, but only for such periods as the vessel is performing the charter service;
- to trade only within agreed limits, to carry only agreed kinds of cargo, and to call only at safe places;

---

3 *The Hill Harmony* [2001] 1 Lloyd's Rep 147, 156.
4 See Coghlin *et al*, *Time Charters* (7th Edn., 2014) (*'Time Charters'*), I.28–I.29.

- to supply the vessel with necessary bunkers at its expense;
- to pay for the costs of calling at the places at which the vessel calls;
- to cooperate in respect of the loading and unloading of cargoes;
- to indemnify the owner against losses suffered as a result of complying with its orders; and
- to redeliver the vessel only at the end of the charter period.

(b) ***Standard-form time charters for oil and gas carriage***

There are numerous standard-form time charters in use for the carriage of oil and gas cargoes. Many are focused on particular trades and address issues specific to those trades. Common examples include Shelltime 4 (used principally for crude oil but also more widely in oil trading), Shell LNGTime1, BPTIME 3, TexacoTime, ExxonMobil Time 2005 and BIMCO's Uniform Time Charter Party for Vessels Carrying Liquified [sic] Gas (GASTIME). Additionally, parties may adapt these or other forms for use for their charter or prepare their own bespoke contract. Given the limited scope of the present analysis, it is not possible to consider the terms of specific agreements in detail. The focus is, therefore, on common provisions fundamental to the carriage of such cargoes.

(c) ***The performing vessel***

A time charter will identify the vessel that it is agreed will perform the charter, although this may be subject to a right to substitute another vessel at any time. The charter will also typically provide comprehensive details as to the vessel's design, carrying capacity, performance (including its speed and its rate of consumption of bunkers), operational history, and, if appropriate, the equipment on board for handling the cargo, including any means for heating, pressurising, cooling, agitating and pumping. The charter will normally also contain, if appropriate, details of any oil major or other approvals and certificates necessary for proper performance of the anticipated carriage. The owner will be responsible for ensuring the accuracy of the description in the charter, generally at the time of conclusion of the charter and often also at the time of delivery.

An owner will similarly be generally responsible for ensuring that the vessel is seaworthy and fit for the service expected during the charter. Again, these obligations may need to be satisfied at the time of conclusion of charter although more typically they are also required to be satisfied at the time of delivery. Whereas obligations as to fitness for service are generally expressly set out in a charter, obligations as to seaworthiness may either be implied or incorporated where the Hague-Visby Rules (or equivalent) form part of the charter.[5] The obligation to deliver a vessel that is seaworthy requires that the "ship must have that degree of fitness which an ordinary careful owner would require his vessel to have at the commencement of her voyage having regard to all the probable circumstances of it."[6] Thus, the scope of the seaworthiness obligation is generally subsumed by the scope of the obligation as to

---

5 See 2.1(h) and (i) below.
6 *FC Bradley & Sons Ltd v Federal Steam Navigation Co Ltd* (1926) 22 Ll L Rep 446.

the fitness of the vessel.[7] Fitness and seaworthiness extend both to all aspects of the vessel's ability to handle the intended cargo and to the competence and qualifications of the crew.

These obligations may variously be absolute or they may be qualified, eg, an obligation to use 'best endeavours' or 'due diligence' only. Those relating to the description of the vessel and/or its condition upon delivery are more apt to be absolute, whereas those relating to fitness or seaworthiness throughout the duration of the charter are more apt to be due diligence obligations only; the standard required will depend on the specific terms of the charter in question. The former standard requires that the vessel be in the stipulated condition, generally without exception other than fault of the charterer. The latter due diligence' standard typically requires an owner to carry out inspections, repairs and other works which a skilled and prudent shipowner would carry out if trading the vessel for its own account, and to ensure that any work in fact carried out is done with reasonable care, skill and competence.[8]

A charter will invariably require that the vessel be "classed", ie, certified as compliant with the conditions and requirements of a stipulated class of vessel by an international classification society (eg, Lloyds, Bureau Veritas, DNV, RINA, Class NK and ABS). Similarly, a charter will require that the vessel comply with all international and national regulatory requirements applicable to vessels of that class and/or involved in that trade (eg, regulations of the International Maritime Organisation, international pollution conventions and national maritime laws). Although again subject to the precise terms of the charter in question, the obligation to comply with these requirements will typically be regarded as absolute as at the time of conclusion of the charter and/or at the time of delivery of the vessel and, as a matter of due diligence, throughout the course of the charter.

Failure by an owner to ensure that the vessel is in the required state will constitute a breach of contract that will entitle the charterer to sue for any properly recoverable losses.[9] A charterer may also, or alternatively, be able to refuse to take delivery of the vessel or to terminate the charter in the event of serious non-compliance by an owner, that is, non-compliance with a fundamental term of the charter or serious non-compliance with a less significant term but one which nevertheless deprives the charterer of substantially the whole benefit of the charter.

The charterer is required to "redeliver" the vessel to the owner at the end of the term of the charter, typically in the same condition as at the time of delivery but with due allowance for fair wear and tear.

(d) *Maintenance of the vessel*

Time charters typically require an owner to maintain the vessel in a fit state in all respects throughout the duration of the charter. The extent of the obligation is generally to exercise due diligence, as an absolute obligation would be difficult to satisfy in practice, particularly so the longer the duration of the charter. Where the

---

7 *Time Charters*, 8.10–8.11.
8 *Ibid*, 8.26.
9 See generally chapter 12, *Damages*.

Hague-Visby Rules (or equivalent) are incorporated into the charter, the owner is under an additional, although largely duplicative, requirement to exercise due diligence to make the vessel seaworthy before and at the beginning of each voyage performed under the charter.

Time charters, particularly in the oil and gas sector, typically include provisions relating to the cleaning and preparation of cargo tank spaces before loading of cargoes; where, however, they do not and/or in any event, an owner will typically be required, in the performance of its maintenance obligations, to exercise due diligence to clean cargo spaces.[10]

Failure by an owner to comply with these obligations may entitle a charterer to sue for recoverable losses. Further or alternatively, the charterer may be entitled to place the vessel off-hire.[11]

(e) The charterer's orders

It is an essential feature of a standard time charter that, following delivery of the vessel, the owner must comply promptly with the charterer's orders as to the employment of the vessel, provided that the orders given are lawful and contractual.

In general terms, but subject to the owner's freedom to agree otherwise, a charterer may only give orders for the carriage of cargoes within the vessel's agreed geographical trading limits. In a typical time charter, the trading limits are generally broadly defined and exclude only those areas dictated by concerns as to the safety of the vessel, the crew or the cargo, the unavailability of insurance coverage, or political considerations. Where, however, a time charter is for performance of a single voyage (ie, a trip time charter) rather than for a period of service, the description of that voyage will set the limits on the charterer's orders.[12]

An owner is entitled to refuse to comply with an order to trade outside the agreed limits. It may be entitled to be indemnified by the charterer against any losses suffered even if it does agree to comply with such an order.[13]

In addition to geographical restrictions, time charters invariably permit a charterer only to order a vessel to loading places (ports, berths, moorings etc.) that are 'safe'. In outline,[14] this restriction obliges a charterer not to order a vessel to a place which, at the time the order is given, is prospectively unsafe. In oil and gas charters, the obligation may, variously, be either absolute or only one to exercise due diligence.[15] 'Safe', in this sense, has a broad definition: a port, for example, will not be safe "unless, in the relevant period of time, the particular ship can reach it, use it and return from it without, in the absence of some abnormal occurrence, being exposed to danger which cannot be avoided by good navigation and seamanship."[16] The concept of safety is not limited to the physical integrity of the vessel; it may

---

10 *The Liepaya* [1999] 1 Lloyd's Rep 649, 668.

11 See 2.1(f) below.

12 *Time Charters*, 5.6.

13 *The Kanchenjunga* [1990] 1 Lloyd's Rep 391, 397.

14 At least as a matter of English law, this is a difficult and complex area of law which cannot be considered in detail here.

15 See, for example, clause 4(b) of Shelltime 4; cf clause 5 of GASTIME. See also *Time Charters*, 37.41.

16 *The Eastern City* [1958] 2 Lloyd's Rep 127, 131.

include the wellbeing of the crew, the integrity of the cargo and the risk of the vessel, crew or cargo being subject to legal proceedings or other grounds for detention.

If a charterer orders a vessel to a place that is unsafe in this sense, then, subject to the particular circumstances of the case, it is liable to be held to be in breach of the charter and, accordingly, liable to the owner for losses arising as a result of proceeding to that place. An owner is entitled to refuse to comply with such an order, although a failure to refuse would not necessarily amount to a waiver by the owner of its rights.

If the place to which the vessel is ordered becomes unsafe after the order is given, then the charterer will not be in breach of the charter for having ordered the vessel there but will, if the owner so requires, be obliged to give fresh orders.[17]

A charterer may only order an owner to carry contractual cargoes. Charters for general trading generally permit the carriage of all cargoes except those expressly excluded. By contrast, charters for specific trades in the oil and gas sectors, either expressly or by necessary implication given the specialised physical characteristics of the vessel, are often limited to the carriage of only one or a small number of closely related products. Charterers are also generally if not invariably prohibited, impliedly if not expressly, from giving orders for the loading of dangerous cargoes. Such a prohibition will also be incorporated where the charter incorporates the Hague-Visby Rules (or equivalent).

Finally, charterers are generally prohibited from giving orders for the carriage of unlawful cargoes, such as those the carriage of which is prohibited under domestic or international laws, including sanctions regimes.[18]

Many oil and gas cargoes are potentially dangerous, either because of their inherent nature (eg, toxicity, flammability) or the conditions under which they are necessarily carried (eg, extreme temperatures, under pressure). Such cargoes are obviously not prohibited, *a fortiori* when expressly or necessarily permitted by the terms of the charter, provided that appropriate and normal precautions can be and are taken, adequate notice of the characteristics of the cargo is given to the owner and the cargo is not exceptional for a cargo of that type.

The prohibition on orders to load non-contractual, unlawful or dangerous cargoes is generally absolute, ie, a charterer cannot rely on the fact that it did not know the true nature of the cargo.[19] An owner is entitled to refuse to load any such cargo and, if loaded, is entitled to sue for any recoverable losses it may have incurred as a result of its carriage.

Subject to these matters, an owner is generally obliged to comply with a time charterer's orders as to the employment or economic utilisation of the vessel.[20] The owner, via the master of the vessel, remains responsible for the implementation of those orders, however, including as to matters of navigation and seamanship.

17 *Time Charters*, 10.1, 10.51–10.52.
18 See chapter 16, *Sanctions*.
19 *The Giannis NK* [1998] 1 Lloyd's Rep. 337, at 344–345.
20 There may be additional types of charterer's orders with which an owner may legitimately not comply, such as to deliver cargo other than in accordance with the terms of bills of lading issued in respect of it: see further chapter 8, *Bills of lading*.

A failure by an owner to comply with a charterer's lawful order is a breach of charter in respect of which a charterer may be entitled to sue for its recoverable losses.

An owner may, however, be entitled in certain circumstances to be indemnified in respect of any adverse consequences of complying with a charterer's orders. The justification for this is that the owner has put its vessel at the disposal of the charterer, who can choose where to send it and what cargoes to carry the *quid pro quo* for the owner agreeing to follow the charterer's orders being that the charterer and not the owner should ordinarily bear the normal consequences of those orders.[21]

A matter of particular interest in the carriage of oil cargoes is whether or not a charterer may give instructions for, or the vessel may conduct, on-board blending of different parcels and/or types of cargo. Provision for this is not typically included in the body of standard form charters but is often the subject of detailed additional rider clauses or *ad hoc* agreements. Absent express provision, a charterer will not usually be entitled to require an owner to blend cargoes on board.

Recent international regulations have prohibited blending during sea voyages, though this may be permissible, subject to local regulations, within port limits. Where blending is permitted and can be carried out safely by the vessel, a charterer will typically undertake to bear responsibility for all costs involved in the blending operation and indemnify the owner for all adverse consequences, including the failure of the blended cargo to achieve the desired specification, contamination of the vessel, especially its tanks, and any claims by third parties arising out of or in relation to the blending operation. Typically, fresh bills of lading will also have to be issued in respect of the 'new' cargo.[22]

(f) Hire and off-hire

In exchange for making its vessel available to the charterer, the owner is entitled to be paid hire. Under a time charter, hire is typically calculated as an amount per day running from the time the vessel is delivered to the charterer until the time it is redelivered by the charterer back to the owner and the charter comes to an end. Hire is typically paid in instalments and in advance, although this is subject to negotiation in each case. It is currently a matter of some dispute whether or not, in the absence of an express contractual provision, a charterer's failure to pay hire, without more, entitles an owner to withdraw its vessel from the charterer's employment and terminate the charter.[23] Time charters typically provide, however, for a contractual right to withdraw the vessel and terminate the charter upon notice being given that that right will be exercised if the failure to pay is not remedied within a short period of time. An owner will be entitled to withdraw the vessel, whether or not such notice is given, and treat the charter as having come to an end, if the charterer's refusal to pay hire is properly characterised as evincing an intention

21 See *Time Charters*, 19.15.

22 See Section 2.1(h) below.

23 Contrasting views have been expressed on this point in recent decisions of the High Court in England & Wales: see *Spar Shipping AS v Grand China Logistics Holding (Group) Co Ltd* [2015] 2 Lloyd's Rep 407 and *The Astra* [2013] 2 Lloyd's Rep 69.

not to perform the charter. This will generally be without prejudice to the parties' other rights, including in respect of any prior breaches of the charter and, in particular, the owner's right to sue for recoverable damages arising by reason of the early termination of the charter.

Hire is earned continuously from the time of delivery to the time of redelivery save where the vessel is 'off-hire'. In outline,[24] a time charterer is generally not obliged to pay hire for such periods of time as when the vessel is not available to perform the service immediately required of it for particular reasons, typically but not exclusively involving mechanical breakdown of or damage to the vessel or some other inefficiency relating to the vessel or the crew, save where the vessel's inability to perform is the result of something for which the charterer is responsible. Time charters, particularly in the oil and gas trades, typically also contain detailed provisions providing for off-hire in the event of delays due to breakdown in the specialised equipment needed to carry such cargoes. In order for the vessel to be off-hire, it is not necessary that the cause of the interruption to the charter be a breach of the charter on the part of the owner.[25] Time charters also typically provide for *pro rata* reductions in the hire payable where the interruption is partial rather than complete, such as where performance of tasks is slowed.

In general terms, therefore, it will be the time charterer that bears the consequences of delays in terms of still being required to pay hire to the owner. The charterer will only be able to reverse this position if it is entitled to place the vessel off-hire in accordance with the relevant provisions of the charter. This means, again in general terms, that the charterer will be liable for delays due to port or berth congestion, terminal inefficiency and non-availability of cargo.

(g) Loading and discharge of cargo

An essential element of a time charter is that the vessel will load and discharge cargo as directed by the charterer. The charter will allocate responsibility, which may be divided as between the parties, for those activities. This allocation may be made expressly in the charter or it may be incorporated by reason of there being a recognised practice in the trade and/or at the place where the cargo operations take place or indeed by reason of the nature of the cargo. In the absence of either express provision and/or relevant practice to the contrary, responsibility will generally fall on the owner.[26] This responsibility includes not only providing such ship-side assistance as may be necessary for the physical process of loading and unloading, in which, in practical terms the owner must necessarily be involved, but also legal and financial responsibility for cargo operations. It may, therefore, cover such matters as the cost of the cargo operations, the consequences of physical or other damage to the cargo, the vessel or persons, and the consequences of negligent performance of cargo operations.

---

24 This is, again, a detailed and complicated area, at least as a matter of English law.
25 *The Ioanna* [1985] 2 Lloyd's Rep 164, 167.
26 *Time Charters*, 20.1, 20A.1.

(h) *Bills of lading*

A time charter will typically require the owner to issue bills of lading in respect of the cargo loaded on board the vessel as directed by the charterer. The nature and effect of these instruments is considered in more detail in a later chapter.[27] In general terms, however, the effect of the owner issuing a bill of lading is generally to bind it contractually to the cargo owner (or subsequent transferees of title to the cargo) to take proper care for the cargo and to deliver it at the discharge port only to or as directed by the cargo owner. Time charters will typically contain a term (often referred to as a 'clause paramount' or 'general paramount clause') requiring all bills of lading to be subject to the terms of the Hague Rules,[28] the Hague-Visby Rules[29] or the Hamburg Rules,[30] or to national legislation incorporating those rules into domestic law.[31] These rules contain, among other things, a detailed code covering the owner's handling of and care for the cargo during a voyage. A time charter will also typically include an express or implied undertaking on the part of the charterer to indemnify the owner in respect of the consequences of carrying goods pursuant to a bill of lading which imposes obligations on the owner that are inconsistent with those arising under the charter.

(i) *Care for cargo*

An owner is obliged to take reasonable and proper care for the cargo to the charterer under the charter and, typically, to the owner of the cargo under the bill(s) of lading. Time charters may contain stipulations as to how the cargo is to be handled, including at what temperatures and/or pressures it is to be maintained. Those stipulations may be detailed, but more frequently they require the vessel to comply strictly with all instructions given by or on behalf of the charterer.

As previously noted, a time charter will also typically provide for the incorporation of the terms of the Hague, Hague-Visby or Hamburg rules into any bills of lading issued in respect of the cargo. Typically, although not invariably, a time charter will also incorporate the terms of those rules into the charter itself so as to regulate the position in respect of care for the cargo as between owner and charterer. Where, however, any of those rules are incorporated into a charter, their effect will have to be considered alongside any relevant express terms in the charter itself.

In outline,[32] the Hague-Visby Rules provide that an owner shall exercise due diligence to make its vessel seaworthy at the beginning of each voyage (article III(1)) and shall "properly and carefully load, handle, stow, carry, keep, care for, and discharge the goods carried" (article III(2)). The latter obligation is, however, subject

---

27 Chapter 8, *Bills of lading*.

28 International Convention for the Unification of Certain Rules of Law relating to Bills of Lading 1924.

29 The Hague Rules as amended by the Protocol to Amend the International Convention for the Unification of Certain Rules of Law relating to Bills of Lading 1968.

30 United Nations Convention on the Carriage of Goods by Sea 1978.

31 For example, Carriage of Goods by Sea Act 1979 (UK); Carriage of Goods by Sea Act 1936 (USA).

32 The Hague Rules and the Hague-Visby Rules are, for present purposes, materially identical. A detailed consideration of the terms of the rules is beyond the scope of this chapter. (For such, see Cooke *et al*, *Voyage Charters* (4th Edn, 2014), chapter 85 (*'Voyage Charters'*); Aikens *et al*, *Bills of Lading* (2006), 10.201-10.260.) So too is consideration of the Hamburg Rules, which are, in practice, far less regularly incorporated into charters in the oil and gas trades.

to article IV. Article IV(1) excludes liability for loss arising from the vessel's unseaworthiness unless due to a want of due diligence on the part of the owner in making the vessel seaworthy as required by article III(1). Article IV(2) sets out a number of exceptions to an owner's liability for damage or loss to goods carried on board its vessel. These include "act, neglect or default of the master ... or servants of the carrier in the navigation or in the management of the ship", fire (unless caused by actual fault or privity of the carrier), perils of the sea, strikes and riots, "wastage in bulk or weight or any other loss or damage arising from inherent defect, quality or vice of the goods", "latent defects not discoverable by due diligence", and "any other cause arising without the actual fault or privity of the carrier".

(j) *Exceptions*

A time charter will also typically contain provisions exempting the owner from liability more generally, ie, not only in respect of cargo loss or damage as provided for where the Hague-Visby Rules (or equivalent) are incorporated. These provisions can be differently worded and bear close scrutiny in each case. However, they typically exempt liability for loss, damage or delay on grounds similar to, but not as extensive as, those set out in article IV(2) of the Hague-Visby Rules. Thus, they typically exempt liability for negligence on the part of the master or crew in the navigation or management of the vessel, fire, accidents at sea (eg, collisions, strandings), latent defects in hull, machinery or equipment, extreme weather or natural phenomena, acts of war, riots and strikes.

2.2 Voyage charters

Many of the key features of voyage charters mirror those considered above in the context of time charters. They can, therefore, be discussed more briefly.

(a) *Key features*

In substance, a voyage charter is a contract pursuant to which the owner of the vessel agrees to carry a specific cargo or cargoes on one or more predetermined voyages in exchange for payment of freight. The owner remains in possession of and retains full authority over the vessel; it exploits the vessel's earning capacity itself and thus bears the commercial risk but enjoys the full benefit of the net earnings of the vessel.[33] An owner generally undertakes:

- to present the vessel as described in the charter;
- to arrive at the designated loading port or other place when required by the contract and, upon arrival, to give proper notice of readiness to load the charterer's cargo;
- to ensure that the cargo spaces are ready to accept cargo;
- to load the cargo and subsequently to discharge it at the designated discharge port or place;
- to meet all operational expenses; and
- to issue bills of lading as directed by the charterer.

---

33 *The Hill Harmony* (note 3 above), p 156.

The charterer typically undertakes:

- to provide and load a full and complete cargo as stipulated in the charter;
- to pay freight in respect of the full contractual quantity of cargo, or deadfreight in respect of any shortage;
- to pay demurrage, in the event that loading or unloading takes longer than permitted; and
- to indemnify the owner in respect of losses arising from compliance with its orders.

(b) Standard form voyage charters for oil and gas carriage

There are numerous standard-form voyage charters in use for the carriage of oil and gas cargoes. Arguably the most significant in the oil trade is ASBATANKVOY (1977) (formerly EXXONVOY (1979)). Other significant standard forms are SHELLVOY 6 (2005), BPVOY 4 (1998) and EXXONVOY (1990). For gas cargoes, there is BIMCO's GASVOY (2005) and the International Group of Liquefied Natural Gas Importers' LNGVOY (2012). Notwithstanding this variety, the analysis which follows seeks once again to focus on common provisions fundamental to the carriage of these cargoes.

(c) The performing vessel

A voyage charter will identify the vessel to perform the contractual voyage, although not infrequently the owner may have a right, subject to the charterer's approval, to nominate a performing vessel at a later date or to substitute an alternative vessel. The charter will also provide a detailed description of the vessel, particularly of its cargo spaces and cargo handling equipment, and will confirm that the vessel has in order all necessary certificates and other documentation to permit it to perform the voyage.

A voyage charter will typically contain an undertaking on the part of the owner that the vessel is fit for service and/or seaworthy as from, at the latest, the date of commencement of the journey towards the loadport. It will also typically require, either expressly and/or through incorporation of the Hague-Visby Rules (or equivalent), that the owner maintain the vessel in such a state and as per the description throughout the currency of the charter.[34]

(d) The cargo

Under a voyage charter, the charterer is typically under an absolute duty to provide cargo for loading, save in very exceptional circumstances.[35] The charter will identify that cargo as well as the quantity to be loaded. The cargo provided, even though it matches the description, must be of such a type and in such a condition as would be considered reasonable in the light of the provisions of the contract.[36] The quantity will typically be expressed as either being a full and complete cargo, in which case the charterer must provide enough cargo to fill all of the vessel's normal cargo spaces,

---

34 As to the content of these requirements, see 2.1(c) and (d) above. See also *Voyage Charters*, 11.18–11.69, 52.4–52.8.

35 *The Nikmary* [2004] 1 Lloyd's Rep 55.

36 *Voyage Charters*, 6.43.

or as a specific amount, in which case the charterer must provide at least that amount. In either case, proper allowance must be made for the propensity of the cargo to expand or contract upon loading and/or during the voyage. Subject to express stipulation to the contrary, the charterer is obliged to ensure that the cargo is properly prepared for shipment. The cargo to be carried must also be safe, in the sense already considered.[37]

(e) *Nomination of loading and discharge places*

A voyage charter will normally state the loading and discharge places at the time it is concluded, although conceivably these details can be confirmed subsequently. A charter may, however, expressly state the latest times by which these details are to be provided. Absent such stipulation, the loading place(s) must be confirmed so as to permit the vessel to meet its cancelling date,[38] whereas the discharge place(s) would likely have to be confirmed by no later than the time for issue of bills of lading which will identify on their face the place(s) of delivery of the cargo.

A voyage charter may contain an express warranty on the part of the charterer as to the safety of the nominated loading or discharge places. In the absence of an express warranty, it is a matter of contractual construction as to whether such a warranty should be implied. An important consideration will be the degree of flexibility the charter accords to the charterer to nominate the place: the more certain the places of loading or discharge are at the time the charter is finalised, the more likely the owner will be held to have satisfied itself that the place is safe or that it was prepared to take the risk that it is safe. Conversely, the more flexibility that is left to the charterer, the more likely it is to be held that the charterer is impliedly warranting that any place it chooses is safe.[39] If the charter does contain a warranty as to the safety of the loading or discharge places, then the position will be broadly the same as in respect of time charters, save that it is not presently clear whether or not the charterer would be obliged to make a fresh nomination in the event that a place that was prospectively safe at the time of nomination subsequently becomes unsafe.[40]

(f) *Laydays and cancellation*

Since it is obliged to have cargo ready and available for loading, a voyage charterer will want to know the time of the vessel's anticipated arrival at the designated loading place and its readiness to load. A voyage charter will, therefore, typically contain both an undertaking on the part of the owner that the vessel will proceed to the loading port "with all convenient speed" or "with due dispatch" and either a window within which ('laydays') or a date by which (the 'cancellation date' or 'laycan') the vessel is required to arrive at the designated loading place, failing which the charterer is entitled, although not obliged, to cancel the charter. A voyage charter will typically set out in detail how the right to cancel is to be exercised. Voyage charters also

---

37 See 2.1(e) above. See also *Voyage Charters*, 6.47–6.63.
38 See 2.2(f) below.
39 See *Voyage Charters*, 5.33.
40 *Ibid*, 5.52–5.56.

typically contain provisions permitting the owner, in the event that it appears likely that the vessel will not arrive by the cancellation date, to call upon the charterer in advance to declare whether or not it will exercise its right to cancel the charter. The charterer's contractual right to cancel is separate from any other right it may have to terminate the charter (eg, because of repudiatory breach on the part of the owner) and it will not itself typically confer a right to damages in the event that the vessel is cancelled. A charterer may, however, be entitled to sue for breach of any undertaking by the owner at the time the charter is concluded as to the location or anticipated readiness of the vessel and/or any failure on the part of the owner to take reasonable steps to permit the vessel to arrive at the loading place by the cancellation date.

(g) ***Laytime and demurrage***[41]

A key feature of a voyage charter is that it affords the charterer defined or determinable amounts of time within which to complete loading and discharge of the cargo or a total amount of time for both sets of cargo operations ('laytime'). The trigger for commencement of running of the permitted time will be the giving of a notice of readiness to load or to discharge by the owner in conformity with the requirements of the charter. A failure on the part of the charterer to load and/or discharge the cargo within the allowed laytime is a breach of contract. Almost invariably, voyage charterers provide for liquidated damages ('demurrage') to be payable by the charterer to the owner in respect of such breach. Some voyage charters provide for the owner to compensate the charterer (by payment of 'despatch') if the allowed time is not fully used.

(h) ***Loading and discharging operations***

A voyage charter will typically identify which party is to be responsible, physically and otherwise, for cargo operations. In the absence of full or complete provision for this, responsibility will be determined by reference to the custom at the port where the operation is being performed and/or the custom in the particular trade. Detailed provisions are, however, common in oil and gas charters given the specialised equipment and procedures necessary for safe handling of cargoes.

(i) ***Freight/deadfreight***

The owner's remuneration for performance of its obligations under a voyage charter is the payment of freight. Freight may be payable on a lump sum basis but more typically, particularly in oil and gas charters, it is calculable by reference to the weight or volume of the cargo shipped. There are often detailed provisions as to how, where and at what time that weight or quantity is to be measured. Freight will typically be deemed earned upon loading and payable upon delivery of the cargo; certain charters, however, make freight payable on other bases.

---

41 This is a detailed and complex area of law and practice and only a very brief outline is given here. Voyage charters typically contain or incorporate detailed provisions covering all or at least some aspects of the commencement and running of laytime and of liability for demurrage, which can vary greatly as between charters. A full treatment of the area is to be found in Schofield, *Laytime and Demurrage* (6th Edn., 2011). See also *Voyage Charters*, chapters 15, 16, 57 and 58.

Where a charterer fails to provide the full contractual amount of cargo, it will be in breach of contract and liable to pay damages to the owner ('deadfreight'). The charter may specify how such damages are to be quantified but, failing this, they will be quantified in accordance with normal principles, ie, so as to put the owner financially in the same position as it would have been in had the charterer not breached the charter.

Oil and gas charters typically provide that, where the cargo supplied by the charterer falls short of the contractual minimum quantity, the owner shall seek confirmation from the charterer as to whether it intends to provide further cargo. If no further cargo is provided, the vessel may then proceed on its voyage subject always to it being in a safe and seaworthy condition.

(j) *Bills of lading*

Like time charters, voyage charters typically require the owner to issue bills of lading in respect of the cargo loaded on board which incorporate the terms of the Hague-Visby Rules (or equivalent). Voyage charters frequently contain an express or implied indemnity on the part of the charterer and in favour of the owner where signing of the bill exposes an owner to liability it would not otherwise have faced under the charter alone.[42] However, voyage charters, particularly in the gas and oil trades, often require the use of a specified or bespoke form of bill of lading, so the risk of there being any material inconsistency in the owner's obligations is minimal.

(k) *Care for cargo*

The position here is broadly the same as in respect of an owner under a time charter.[43] The owner will be under an express and/or implied obligation to take reasonable and proper care for the cargo. Typically, the charter will incorporate the terms of the Hague-Visby Rules (or equivalent) and, in oil and gas trades, charters are also likely to contain detailed provisions dealing with the specific requirements (eg, heating, refrigerating, agitating) for handling of the cargo being carried.

(l) *Exceptions*

The position is similar to that pertaining under time charters.[44]

3. Contracts of affreightment

At least in modern English legal parlance, the term 'contract of affreightment' ('CoA') is generally used to describe an umbrella contract providing for a series of voyage charters for shipments of the same type of cargo between the same places over a defined period of time.[45] CoAs are typically tailor-made to meet the specific requirements of both the carrier and the buyer or shipper of the goods, in particular to:retain a measure of flexibility as to the precise time and quantity of, and the

---

42 See, further, 2.1(h) above; see also *Voyage Charters*, 18.223–18.230.
43 See 2.1(i) above.
44 See 2.1(j) above.
45 Traditionally, the term was used more broadly as a synonym for a contract of or for carriage including both charterparties and bills of lading. This usage is still seen.

identity of the performing vessel for, each shipment, and also fix the costs of shipment in advance for the duration of the CoA.

In these circumstances, standard form terms are rarely used for CoAs.[46] A CoA will, however, typically set out the key commercial terms agreed by the parties, such as:

- the type of cargo to be shipped, the number and frequency of shipments over the term of the CoA;
- the places of loading and unloading;
- the quantity to be shipped, both in aggregate and in each individual shipment; and
- the rates and terms of payment of freight and demurrage.

The terms of a standard form voyage charter or those of a previous voyage charter concluded between the parties will typically be incorporated to provide the detailed provisions governing each shipment once the vessel has arrived at the loadport.

Since it is not generally envisaged that each shipment will be performed by the same vessel, and since it is typically intended that there be some flexibility as to the timing of each shipment (provision to the effect that shipments shall be "fairly evenly spread" over a defined term are common), a CoA will typically also contain detailed provisions for the nomination and confirmation of performing vessels and for establishment of the shipment schedule. These provisions vary greatly between CoAs, including as to which party is to initiate the procedure, the degree of latitude available to the parties, the steps required to complete the procedure, and which party will have the final say.

Less typically, a CoA will specify the consequences of non-performance by one of the parties. Where this is addressed, provision may be made for suspension of performance by the innocent party, abandonment of a particular shipment with or without compensation, or for termination of the CoA. As a general rule, but subject necessarily to the terms of the contract in question and the specific circumstances of the case, English law will treat individual shipments under a CoA as separable voyage charters, with the effect that failure to perform one shipment will not necessarily amount to or justify termination of the CoA as a whole. Non-performance of multiple shipments or, in appropriate circumstances, even just one shipment may, however, justify termination of the CoA.

4. Other modes of carriage

Oil and gas are routinely transported by other modes of carriage, of which the most significant is by pipeline. There are vast networks, both within countries and transnationally, particularly in Western, Eastern and Central Europe, Central Asia, and North America. However, carriage of oil and gas by pipeline raises legal and practical issues that are quite distinct from those arising in relation to carriage of such goods by sea or by other means, such as rail or road. There are also important

46 They do, however, exist: BIMCO, for example, has produced a standard Tanker Contract of Affreightment (INTERCOA 80).

differences in the regimes typically applicable to different hydrocarbon products. It is not possible here to consider the specific regimes applicable to such carriage in any detail.

Significantly, carriage by pipeline, where it is not provided directly by a public authority, is typically tightly regulated by competition authorities in light of the limits on available infrastructure, the potential for abuse of monopolistic market positions and also the public health and environmental hazards posed by this mode of carriage. This has consequences for the form of pipeline carriage contracts. Such contracts also tend to be long-duration period contracts, since certainty of supply and demand and the ability to fix prices are key considerations.

There are a number of broadly-defined types of carriage contract. One of these is 'contract carriage' which involves shippers entering into contracts directly with the owner or manager of the pipeline on such terms as may be agreed but subject to any regulatory restrictions. Those terms will cover essential commercial terms, such as supply to and discharge from the pipeline, price, and quantity or volume commitments, and whether those quantities are firm or interruptible or some combination of the two. They will also typically include terms relating to priority of access, parking or storing of excess quantities, financial penalties for over- or under-supply or demand, and responsibility for lost or contaminated product. Another form of carriage, typical where the carrier is a public body, is 'common carriage'. This involves carriage being provided on uniform terms for all users, with issues of supply and capacity being determined on some equitable basis. Other forms of carriage, such as 'merchant carriage', where the carrier owned the gas, are now less common.

Rail and road carriage, particularly of oil, is significant in many parts of the world. Such carriage is, however, not generally subject to standard form contracts in international usage.

Quality and quantity determination

Jane Fitzgerald
Ince & Co LLP

1. Introduction

This chapter addresses common issues relating to quality and quantity in international sale contracts as well as the determination mechanisms generally adopted in oil and gas contracts. The scope for disputes over whether the seller has delivered goods of the quality and quantity agreed in the contract and, in turn, whether the buyer has got what he paid for is considerable. In the context of the international sale of goods in particular, buyers and sellers (who, by definition, are likely to be based in different geographical locations) face very real practical difficulties when seeking to verify the quality and quantity of the goods which are the subject of their sale contract. The industry has developed various solutions to overcome these practical issues to enable traders to have confidence in the mechanism for establishing whether the goods delivered correspond with the goods contracted for with regard to quality and quantity.

2. Quantity: general considerations

Traditionally it is the bill of lading that has been used to evidence the quantity shipped by the seller and to calculate the price to be paid by the buyer. Almost invariably, however, the more modern practice in contracts for the international sale of oil and gas is to provide for certificates of weight to be issued and for those certificates to contain the relevant quantity information for invoicing purposes. The method to be used in ascertaining the weight of the goods is usually prescribed in fairly general terms, eg, in accordance with 'good standard practice', 'terminal procedures', 'standard measuring and testing procedures in effect', 'recommended industry standard' or 'customary methods, practice and procedure'. Generally, the measurements are to be taken (or verified) by a mutually agreed independent inspector. Alternatively, the seller may be responsible for obtaining inspection certificates, but provision will also be made for the buyer to have the option of appointing his own representative to witness the relevant measurements.

Historically, ship tank ullage readings have been used to ascertain quantity figures, but where the commodity is loaded/discharged directly from/into static shore tanks, the applicable customary method may be to gauge the shore tanks. Nowadays, terminal facilities often have the capability to measure the quantity by taking flow meter recordings in the loading lines. What is important is that the parties to the sale contract are agreed as to which figures are applicable for the determination of the price. It is particularly important when in the middle of a chain (and acting as both buyer and seller), not to be caught out buying on ullage figures and selling on flow meter figures, or vice versa.

Unless otherwise agreed in the contract, as a matter of English law, the time and place where the quantity is to conform to the contract is 'on delivery'. For sales on shipment terms (eg, classic CIF ('cost, insurance and freight') and FOB ('free on board') contracts), this means that the quantity is to be correct on shipment/loading and for sales on delivered terms (eg, ex-ship contracts) on arrival/discharge. After the applicable delivery point, the risk of the loss of the product will generally transfer from seller to buyer. Particularly in the oil and gas industry, however, parties often seek to modify shipment terms. One such modification, which is peculiar to oil and gas deals, is to transform a classic CIF contract into a 'CIF delivered' contract. On CIF delivered terms, the seller commits that the ship will arrive at the discharge port in time to discharge the cargo within a specified delivery window, notwithstanding the fact that delivery in the legal sense (the trigger for the passing of risk) still occurs on shipment. In a CIF delivered sale, failure to arrive at the discharge port in time to discharge the cargo by the end of the delivery window specified in the contract would be a breach of condition, entitling the buyer to reject, but the quality is still assessed at the time of shipment.

Another adjustment to a classic CIF contract that arises from time to time is an agreement between the parties that, although risk passes on shipment, quantity is to be assessed on an 'outturn' basis, ie, it is the quantity delivered at the discharge port that is relevant. In such contracts, the buyer does not pay against the quantity shown as shipped on the bill of lading or by reference to the load port certificates, but instead on the basis of the actual quantity received at the discharge port. There is often a question mark over the effect of such clauses and specifically whether they operate simply as a price adjustment mechanism or whether they have a broader effect on the allocation of risk between the parties. The prevailing view is that these clauses do not change the fundamental allocation of risk between the parties to a CIF sale, with the result that risk of loss of or damage to the goods during transit remains with the buyer as in a traditional CIF contract, but that the seller can only expect to be paid for the quantity that survives the journey. In other words, in modifying the CIF terms, the parties agree that during transit the risk of loss of quantity remains with the seller.

What happens in the case of a total loss? In *Soon Hua Seng Co Ltd v Glencore Grain Ltd*[1], the relevant C&F contract contained both the provision that risk from warehouse to warehouse was with the buyer and an outturn clause that referred to price being calculated on the basis of the "full outturn weight at destination". In the event, the vessel ran aground and the entire cargo of rice was lost before reaching the intended final destination. Accordingly, the full outturn weight at destination was ostensibly zero and, on a strict interpretation of the pricing clause, the sum payable by the buyer would also therefore be zero. The court held, however, that in circumstances where the rice did not arrive at all, the risk of loss was on the buyer:

> *...[T]he contracts must be viewed as essentially c. & f. in their nature, with the simple proviso that, if and when the goods do arrive and outturn, they will be weighed and the price will in that event be adjusted by reference to the outturn weight. If the goods*

1 [1996] 1 Lloyd's Rep 398.

covered by the shipping documents and invoice are lost in transit and do not arrive at all, the risk of loss remains on the buyers and no question of any adjustment to the payment due against the commercial invoice can arise. (Per Mance J at 405.)

Accordingly, it seems that an outturn agreement is simply an agreement that the buyer just pays for what is actually delivered, so that transit loss is with the seller unless there is a total loss, in which case the risk is with the buyer. Clearly, there is scope for argument about where the line should be drawn. If there is a total loss, that is the buyer's problem, but what if 99% of the goods arrive? Increasingly, in order to avoid any misunderstanding, industry standard general terms and conditions (GT&Cs) seek to address this issue: see, for example, Statoil's 2011 GT&Cs, which provide as follows:

3.2.6 For CIF Outturn and CFR Outturn deliveries, the following terms shall apply:

(a) For the purpose of determining the compliance of the Product with the quantity and quality provisions of the Special Conditions, the quality shall be determined at the Port of Loading...and the quantity measurement shall be carried out at the Port of Destination ...;

(b) In the event of a total loss of the Cargo or where the discharge quantity determined by the Independent Inspector ... is less than 99.5% of the loaded quantity ... then the quantity invoiced by the Seller shall be 99.5% of the quantity specified on the certificate(s) of quantity (or the equivalent document) issued at the Port of Loading.

3. Quantity: remedies

The seller has a duty to deliver the quantity of goods identified in the contract. If the seller delivers the wrong quantity, the buyer is usually entitled to reject the goods. Section 30(1)–(3) of the Sale of Goods Act 1979 (the 1979 Act) sets out the options available to the buyer in such circumstances.

Section 30(1) of the 1979 Act addresses insufficient delivery and permits the buyer to reject the lesser quantity that the seller has sought to deliver or, if the buyer wishes to take the goods, permits the buyer to keep the insufficient quantity delivered (at the contract price). Usually, in commodity sales, the buyer does not pay for the goods in advance and, in that scenario, the buyer would simply pay for the quantity taken. If, however, the buyer has paid in advance, it is entitled to claim back from the seller the balance of the price it has paid for the full cargo. Section 30(2) and (3) of the 1979 Act are the equivalent provisions in respect of excessive delivery and allow the buyer to reject the entire delivery, accept the contractual quantity and reject the excess, or accept the entire delivery including the excess and pay for it at the contract price. The strict duty imposed on the seller is, however, subject to de minimis principles – in other words, slight variations in the quantity delivered will not entitle rejection (these principles operate at common law and by operation of statute by virtue of section 30(2A) of the 1979 Act).

In oil and gas contracts, this strict duty is very often relaxed by stating that the contractual quantity is an approximation or, most commonly, by the inclusion of an expressly accepted tolerance. A tolerance provision allows one or other party the option to deviate from the exact contractual quantity by a certain margin either way, eg, "30,000 mt +/- 10% in seller's option". Leeway provisions such as this considerably reduce the scope for disputes regarding quantity.

4. Quality: general considerations

Provisions as to the quality of the goods sold are usually, principally set out in the contract specification. These specifications set out the required characteristics of the goods by reference to relevant chemical and physical criteria.

As a matter of English law, there is an implied term in a contract for the sale of goods by description that the goods will correspond to that description (section 13 of the 1979 Act). That implied term is a condition of the contract, breach of which entitles the buyer to reject the goods, terminate the contract and sue for damages. The principle is that there is a fundamental obligation on the seller to supply what was described as being for sale. Accordingly, if the seller tenders something other than what was described, the buyer may elect not to take it on the basis that it was not what the buyer bargained for. That is so even if the item actually tendered is in fact better or no worse than what had been described:

> *...Ultimately the test is whether the buyer could fairly and reasonably refuse to accept the physical goods proffered to him on the ground that their failure to correspond with that part of what was said about them in the contract makes them goods of a different kind from those he had agreed to buy...*[2]

The seller's failure to supply goods which comply with the identified quality requirements on the other hand will not necessarily entitle the buyer to reject. There may be situations where it is difficult to distinguish a contractual requirement which is a matter of description (in that it goes to the identity of the goods) from one which is a matter of quality and indeed there may be provisions which can properly be said to be both. Usually however, the specification will be considered to relate to quality rather than description. In an oil context, this issue recently arose in *Proton Energy Group SA v Orlen Lietuva*,[3] where a contract for the sale of oil stated at Clause 3 (headed "Product") that the product being sold was "Oil Blend...CN 2710", and, at Clause 4 (headed "Quality"), that it incorporated the oil specification. In that case, Orlen argued that there were a wide variety of products that could be described as "Crude Oil Mix" and, as such, as a matter of ordinary language and sense, the specification was a matter of description. The judge rejected that analysis and determined that Clause 3 contained the description of the product and Clause 4 dealt with the quality of the product, such that the specification was not part of the description of the goods.

Description will usually be limited to the identification of the product, as such, whether the seller's failure to comply with an aspect of the contractual specification entitles the buyer to reject the goods or not depends upon the nature of the quality term, ie, whether it is a condition, warranty or an innominate term. This issue was recently considered by the High Court in *RG Grain Trade LLP (UK) v Feed Factors International Ltd*,[4] in which Hamblen J confirmed that the relevant legal principles are as follows:

> *Whether there is a right of rejection depends on whether the specification as to quality is a warranty (in which case there will be no right of rejection), whether it is a condition*

---

2 *Ashington Piggeries Ltd v Christopher Hill Ltd* [1972] AC 441 at 503, per Lord Diplock.
3 [2013] EWHC 2872 (Comm).
4 [2011] EWHC 1889 (Comm).

(in which case there will be a right of rejection), or whether it is an innominate term (in which case there will be a right of rejection if the breach of it would deprive the other party of substantially the whole benefit of the contract).

The applicable classification will, of course, depend upon the specific terms of the contract in question but, generally speaking, the requirements of a quality specification are likely to be classified as innominate terms. Accordingly, breach of the specification will usually only entitle the buyer to reject in circumstances where the breach deprives the buyer of substantially the whole benefit of the contract. This will not usually be the case, for example, where one parameter of the spec is off.

A further source of requirements as to the quality to which the goods must adhere is the 1979 Act, which contains various implied terms as to quality applicable to all sale contracts governed by English law, unless expressly excluded by the parties. Section 14(2) implies a condition that the goods supplied will be of satisfactory quality and section 14(3) that they will be fit for the known purpose. Accordingly, the buyer may be able to argue that the breach of the contractual specification renders the goods unsatisfactory or not fit for purpose, each of which are conditions of the contract, breach of which entitles the buyer to reject the goods, terminate the contract and sue for damages.

Very often in commodities contracts, however (and particularly in the oil and gas industry), the parties expressly exclude the terms as to quality implied by the 1979 Act. While different rules may apply in consumer contracts, in business to business contracts it is perfectly possible, as a matter of English law, to exclude the implied terms as to quality. As with all exclusion clauses, however, the wording must be clear: the general principle being that they will be interpreted strictly against the party seeking to rely on the exclusion. This approach reflects the fact that the English courts will not readily accept that a party intended to forgo rights it would otherwise have had as a matter of law, unless that intention is very clearly spelled out in the contract. An example of the strict approach is found in the Court of Appeal decision in *KG Bominflot Bunkergesellschaft für Mineraloele mbH & Co v Petroplus Marketing AG, The Mercini Lady*.[5] In that case, a contract for the sale of gas oil contained an exclusion clause excluding "guarantees, warranties or representations, express or implied, [of] merchantability, fitness or suitability of the oil for any particular purpose or otherwise". Notably, however, the exclusionary wording did not expressly mention 'conditions', which the 1979 Act's implied terms as to quality are (see section 14(6)). As such, the Court of Appeal held that the exclusion clause did not extend to the conditions implied by the 1979 Act; the sellers therefore could not escape those implied statutory obligations of satisfactory quality and fitness for purpose. More recently, however, in considering the following exclusionary wording – "any implied warranty of merchantability or of fitness for a particular purpose" – which once again made no reference to the word 'condition', the court held that, on the particular wording, the clause was sufficient to exclude the 1979 Act's implied obligations as to quality. In the case in question (*Air Transworld Ltd v Bombardier Inc*[6]),

5 [2010] EWCA Civ 1145.
6 [2012] EWHC 243 (Comm).

the judge managed to distinguish *The Mercini Lady* on the facts but, in order to avoid any uncertainty when contracting out of the 1979 Act's implied terms as to quality, express reference to 'conditions' is recommended. In practice, the relevant exclusions are often incorporated into the sale contract by reference to one or other of the industry standard GT&Cs. For example, Shell's GT&Cs for Sales and Purchases of Crude Oil (2010 edition) contains the following provision addressing the issue:

> *28.1.2…This sub-section constitutes the whole of the Seller's obligations with respect to the description, quality and fitness for purpose of the crude oil to be delivered and (save to the extent that exclusion thereof is not permitted or is ineffective by operation of law) all statutory or other conditions or warranties, express or implied, with respect to the description or satisfactory quality of the crude oil or its fitness for any particular purpose or otherwise are hereby excluded.*

5. Quality: remedies

Section 53(2) and (3) of the 1979 Act set out the remedy for breach of warranty of quality as follows:

> *(2) The measure of damages for breach of warranty is the estimated loss directly and naturally resulting, in the ordinary course of events, from the breach of warranty.*
>
> *(3) In case of breach of warranty of quality such loss is* prima facie *the difference between the value of the goods at the time of delivery to the buyer and the value they would have had if they had fulfilled the warranty.*

As is clear from the wording of the 1979 Act, the general rule is that the buyer is entitled to the difference in value between sound and defective goods. However, it is important to note that the difference between sound and unsound value is not always the only appropriate measure of damages and, importantly, it does not operate as limit on the available damages. This has recently been reconfirmed by the court in the context of an appeal against a Federation of Oils, Seeds and Fats Associations (FOSFA) arbitral award: *Saipol SA v Inerco Trade SA.*[7] In that case, a cargo of Ukrainian crude sunflower oil had been contaminated prior to loading and the FOB buyer brought an action under sections 13 and 14 of the 1979 Act, seeking damages not only for the difference between the sound and unsound value of the contaminated sunflower oil but also storage/financing costs and sums paid to the buyer's own sub-buyers due to the contamination. The FOSFA tribunal held that the buyer was only entitled to damages in respect of the difference between the value of the sound and contaminated goods and not the consequential losses resulting from storage and financing charges and sums paid to sub-purchasers to whom contaminated oil had been sold. On appeal from the award, the court had no hesitation in holding that section 53 does not limit recoverable damages to sound versus unsound value since, under section 53(2), there can (in appropriate circumstances) be a claim for consequential losses on the basis that such losses are those which will result in the usual course of events.

---

7 [2014] EWHC 2211 (Comm).

6. Quality/quantity determination clauses

Many commodity contracts contain quality/quantity determination clauses under which the parties agree that a third party inspector will examine the goods and provide the parties with his findings as to quality/quantity in a surveyor's certificate. Such certificates, of themselves, provide good evidence of what has been shipped (and indeed often form part of the documents which are required under the sale contract and are therefore also important in the context of obtaining payment under a letter of credit). It is, however, very common in the oil and gas industry for parties to go one step further and agree that independent surveyor's certificates will be final and binding on them ('certificate final and binding clauses'). These conclusive determination provisions have a number of advantages for traders in terms of efficiency, certainty, finality and the avoidance of disputes which, by their nature, are likely to be evidentially difficult. Due to the nature of the task, measuring quality and quantity of cargoes of oil and gas is inherently uncertain. As advanced and sophisticated as the test methods may be, there will always be limits on the precision achievable in terms of repeatability and reproducibility. In agreeing to a certificate final and binding clause, parties seek to impose some measure of commercial certainty which, albeit somewhat artificial, is efficient and fair to both parties.

The approach of the English courts has generally been to uphold certificate final and binding clauses as valid and effective if their wording is clear. Accordingly, even where the inspector has made a mistake (which is admitted to have been a mistake), the courts have held that the parties could not go behind the certificate in circumstances where the parties had agreed that "official certificates of inspection to be final as to quality".[8] The rationale is that the commercial purpose of such clauses is to avoid disputes, which purpose would be defeated if the courts did not enforce the parties' agreement to stand by the determination of the independent inspector.

In agreeing to be bound by the findings of a third party, the parties are placing considerable trust in the inspector and are each therefore concerned to ensure that the surveyor appointed is reputable. It is obviously important that the inspector is independent, ie, a disinterested third party. The mechanism for the appointment of an independent inspector is usually expressly identified in the sale contract. The contract may empower one or other party actually to take steps to appoint the inspector but, given the potential importance of the inspector's findings, very often provision is made for the parties to make a joint appointment or to appoint a 'mutually agreeable' inspector. This in itself can cause difficulties if the parties cannot reach agreement on a suitable inspection agency. As such, parties sometimes address this issue up front by identifying in their contracts a number of alternative surveying companies, which are in effect 'pre-approved' inspection agencies from which pool the inspector must be selected by the party tasked with appointing the inspector.

Another matter of considerable importance where the certificate is to be final and binding is that of the location of the sampling. The objective is plainly that the samples to be analysed by the inspectors are to be representative of the cargo's

---

8 *Alfred C Toepfer v Continental Grain Co* [1974] 1 Lloyd's Rep 11.

quality on delivery. Since taking drip samples from the loading line is not always practicable, the sampling options tend to be shore tank versus ship's tank. Given the potential for contamination issues to arise in between shore tank and ship's tank (in the line, in the vessel's holds etc) it is important to specify in the contract which location is the relevant one for the inspector's determination. Once again, the industry is alive to the potential for disputes in this regard and some of the most prevalent GT&Cs make provision to bridge the gap between shore tank and manifold. For example, BP's 2007 GT&Cs for Petroleum Products provide as follows:

7.4 Place of Certification

Should it not be customary practice at the Loading Terminal at the time of shipment for measurement and sampling... to take place at the Vessel's manifold immediately prior to loading... then it is a condition of the Agreement that the Seller shall be obliged to provide the same quantity and quality of the Product at the Vessel's permanent hose connection as set out in the certificates of quantity and quality so issued.

In light of the importance to the parties of the determination by the independent inspector, the inspector entrusted with the task cannot delegate that responsibility to another inspection agency. A similarly strict approach applies to the inspector's adherence to instructions. In *Veba Oil Supply & Trading GmbH v Petrotrade Inc*,[9] the quality determination clause provided:

To be determined by a mutually agreed independent inspector at the loading installation in the manner customary at such installations such determination shall be final and binding for both parties save fraud or manifest error.

The quality specification in the contract provided as follows:

Product/Quality Gasoil meeting the following guaranteed specifications:

Test Limit Method ASTM

Density at 15 deg. C + 0.876kg/l max D1298

In the event, however, the independent inspector used a different test method, D4052. The Court of Appeal held that the inspector was required to determine the quality of the gasoil using test method D1298; the inspector had not done so and the determination was not therefore binding on the parties. That decision is not, of itself, particularly controversial as the court was merely upholding the parties' bargain. The controversy came about from the fact that it was common ground between the parties in that case that the test method actually used, ie, D4052 (i) was more modern and more accurate, (ii) was the method customarily used at Antwerp (which was the load port), and (iii) if the independent inspector had used test D1298, he would inevitably have made the same finding as to density as he had done using D4052. Nevertheless, the Court of Appeal held that the determination was not binding because of the inspector's material departure from the instructions. Again, the rationale is understandable: the parties ought not to be bound by the findings of the inspector to the extent that he had deviated from the task which was agreed by the parties.

There is no standard wording for a certificate final and binding clause, but it is very common for such a clause to include a saving provision to make clear that the

---

9 [2001] EWCA Civ 1832.

certificate will not be final and binding on the parties in the case of fraud or manifest error. The principle behind the fraud exception is obvious: the parties do not agree to be bound where the inspector is fraudulent in his assessment of the quality or quantity of the goods set out in the certificate or if the certificate is otherwise the product of fraud. In the parlance of the English courts, "fraud unravels all". The manifest error exception, on the other hand, is slightly more nuanced. Taking the *Veba Oil* example previously discussed, one might think that the use of test D4052 rather than test D1298 might be just the sort of error the parties had in mind when including the saving provision in the wording of their contract. The approach of the courts is, however, to draw a distinction between:

- mistakes (which will not impact upon the finality of the inspector's determination);
- departures from instructions by the inspector (which would render the certificate not binding, since the parties have only agreed to be bound by a determination based on the agreed method); and
- manifest errors so obvious as to be beyond argument (which would provide grounds for challenging the finality of the determination).

Despite how commonly the phraseology appears in certificate final and binding clauses, there is very limited authority on the meaning of the term 'manifest error'. In *Veba Oil* itself, Simon Brown LJ defined it as constituting "oversights and blunders so obvious and obviously capable of affecting the determination as to admit no difference of opinion". Having defined 'manifest error' in that way, the judge went on to indicate that, although on his analysis the question did not arise, even if he was wrong in finding that using the wrong test method was a material departure from instructions rendering the determination not binding, he would not conclude that the use of the wrong test nevertheless constituted a manifest error. Simon Brown LJ's definition has been approved in subsequent cases, with the result that there is very little scope for challenging a determination on this basis:

> *Manifest is a word which gives a very limited window of opportunity to challenge. The examples given in the various authorities above show that it is something like an arithmetical error, or a reference to a non-existent building and the like.*[10]

10 *Walton Homes Ltd v Staffordshire County Council* [2013] EWHC 2554 (Ch).

Practical aspects of quantity and quality determination in relation to shortage and contamination claims

John Minton
Mike Birchall
Richard Minton
Minton, Treharne & Davies Ltd

1. Introduction

Oil and gas trading inevitably involves a sale and purchase contract agreed between at least one buyer and one seller. Fundamental to the function of this process is the proper determination and agreement of the quantity and quality of the goods in question.

1.1 Quantity determination

The quantity of the goods bought and sold is likely to be specified in the contract in terms of an amount plus or minus a certain percentage, the variable quantity being defined as at either the buyer or the seller's option. Quantity is typically measured either by determining the change of volume contained in shore, ship, road or rail tanks between start and finish of transfer operations or by measuring volume flowing through the transfer pipeline with a flow meter. Integral to either process is the determination of the temperature and density of the oil. The latter entails taking a representative sample of the goods for testing.

1.2 Quality determination

The agreed quality of the goods may be specified in relation to either:

(a) a published 'standard' specification – for example, a national or international specification for goods of the type in question. Such a specification will set out a list of required quality criteria, the test methods by which they should be measured and acceptable 'on-specification' limits (usually in terms of maxima or minima) for the results obtained when a representative sample of the goods is tested by the required test methods;

or

(b) a bespoke specification agreed between the buyers and sellers specific to the goods. This is likely to function in the same way as a published standard specification (and may include such), in that it will have a list of required quality criteria and test result limits (maxima and minima) in relation to those criteria.

It should also specify the test methods to be used, but this is not invariably the case. Failure to specify the applicable test method in a specification can be an important omission, as determination of the same criteria by different methods is likely to give rise to different answers being obtained with the potential for dispute. This is discussed further below;

or

(c) a list of 'typical' test results in relation to the goods with limits (maxima or minima) that must not be breached. Again, test methods should be specified but often are not, or are only defined very loosely;

or (very occasionally)

(d) a list of actual test results (again, with or without test method identification).

Broadly speaking, the likelihood of a dispute arising increases the lower the procedure adopted appears in the above list, that is from (a) to (d).

1.3 Disputes

A dispute or claim may arise if there is any significant discrepancy between the seller's and buyer's measure of quantity and/or quality. Various steps are taken to avoid such disputes. For example, for some shipments the parties to a contract may agree that load port shore results or, alternatively, discharge port shore (out-turn) results are final and binding on both sides. They will commonly agree to appoint on a shared basis one independent inspection company to determine an 'agreed' set of quantity and quality results between them. Despite such agreements, disputes still often arise.

This chapter discusses the main issues that can give rise to quantity and/or quality disputes.

1.4 The role of the independent inspector

For traded cargoes, determination of quantity and quality are almost invariably assigned to independent inspection companies, firms set up to provide certification of quantity and quality for their clients. There are a large number of such companies operating around the world, ranging in size from those with only one or two employees and operating at only one port or in one country, through to very large multinational organisations with thousands of staff who can provide their services almost anywhere.

It is the often the case that one inspection company will be appointed on a 50/50 basis, representing both buyer and seller, to monitor the custody transfer of the goods. In the case of multi-stage transfers, such as shipment by sea from a load port to a discharge port, an inspection company, sometimes the same one, will need to be appointed for each stage.

Although this work is assigned to inspectors employed by third party inspection companies, it is necessarily the case that they rely on the equipment of others. For quantity determination, a principal example of this is that they will be reliant on the calibration of the terminal's shore tanks (depth of contents against volume) or flow meters. For quality determination, although in some ports the inspectors will test the samples in their own laboratories, on many occasions they will simply witness and certify tests performed in the shipper's laboratories.

Guidance notes for independent inspectors are issued by the United Kingdom's Energy Institute (EI).[1] Other bodies, such as the International Federation of Inspection Agencies (IFIA), run inspection certification schemes.[2]

1.5 'Super' inspectors and surveyors

In addition to the appointment of the shared inspection company, occasionally either or both parties may appoint an additional technical firm, sometimes known as expeditors or 'super' inspectors, specifically to represent them, perhaps with the aim of influencing events to their advantage, or at least so as not to be put at a disadvantage. The ship owner, assuming the cargo is to be shipped by sea, will not normally be represented other than by the vessel's crew. In the event that something goes wrong during operations, however, the ship owner and the other parties are likely to appoint additional specialist surveyors to investigate matters on their and their insurers' behalf. In such circumstances, it is not unusual to have, say, six different independent organisations all investigating at once.

2. Quantity determination – shortage claims

The quantity of a traded cargo is usually too large for it to be measured by direct measurement such as weighing over a weighbridge, although this does happen from time to time. Instead it is measured indirectly by determining the volume transferred from or into a calibrated shore or ship's tanks and correcting this result to volume at a standard temperature (usually either 15°C or 60°F) using correction factors derived taking into account the measured density and temperature of the oil.[3] Alternatively, loaded quantities (very rarely discharged quantities) can be determined by measuring the volume that has flowed through a calibrated 'in pipeline' flow meter. The density and temperature of the oil are again needed to calculate the quantity but are often determined automatically.

Of course, if the measurements made by the different parties involved in the custody transfer of the traded goods (for example the shipper, the ship and/or the receiver) indicate receipt of a lesser quantity than reportedly shipped, taking into account inevitable losses, then a shortage claim is likely to arise. There are a number of reasons why such shortage claims occur. These broadly fall into two main categories – real (physical) & apparent (paper) losses. These are discussed below.

2.1 Physical losses

Physical losses of oil commonly arise to some extent whenever oil is moved between two locations. Some of these losses are inevitable due to the nature of the cargo and the transfer process employed, and hence will arise despite the oil being properly handled.

---

1 For example, HM29. *Procedures for petroleum product cargo measurements by cargo inspectors.*
2 IFIA Petroleum Inspectors Certification Program.
3 Note: Crude oils and petroleum products expand and contract significantly with changing temperature.

(a) ***Inevitable losses***

Some physical losses are inevitable. The extent of the inevitable loss arising is not easy to determine and so can give rise to dispute. A valuable source of published information available to address this question is a continuing study undertaken over the last 20 years by the EI's Marine Oil Transportation Database Committee, the results of which are published annually in the Institute's journal[4] under the title *Marine crude oil transport – global voyage losses.*

The 2014 Edition of the EI paper shows the results of study of over 6,000 seagoing ship and 1,200 inland barge carriages of crude oil cargoes in 2013 from terminal to terminal or, as often described, shore tank to shore tank. The results of this study show an average NSV (net standard volume = measured volume corrected to standard temperature, net of water and sediment content) loss of 0.17%. This figure must both have a strong relation to the inevitable loss arising in relation to crude oil shipments, although it must also be to some extent influenced by things that have gone wrong in relation to individual voyages, resulting in gains or losses.

In fact this study showed that the standard deviation of the results was 0.31%. By definition, assuming normal distribution, 66% of all results will lie within one standard deviation and 95% within two standard deviations, either side of their mean.

That said, consideration of the EI paper shows that there are some crude oils in relation to which better results than the norm are experienced and others that are worse. These may merit special consideration.

There is no similar database in relation to petroleum products. In the authors' experience, however, the results to be expected for inevitable loss are very similar to those discussed for crude oil shipments, although not necessarily arising by exactly the same causes in the same proportion.

On this basis and returning to the results of the EI study, the mean loss and standard deviation results reported (0.17% and 0.31% respectively) imply that 34% of all voyages will show quantity gains or losses outside the range +0.14 to -0.48%. This is considered to provide good justification for the common 'trade allowance' of 0.5%, as applied in many trading contracts and cargo insurance policies as a deductible in relation to shortage claims. There have been a number of occasions on which a lower deductible has been applied, but in general these are considered misguided by the authors.

For both crude oil and petroleum products, sources of physical loss are discussed below, as well as ways that these can be increased by incorrect handling of the cargoes concerned. Claims can arise when such incorrect handling increases the physical losses above the inevitable.

(b) ***Causes of physical loss – inevitable and otherwise***

Evaporation loss: Volatile cargoes (most crude oils, naphtha and gasoline) will suffer evaporation loss during movement, particularly if the vapours, are vented to the

---

4 *Petroleum Review.* Latest edition published September 2015 stating average NSV loss of 0.18% for shipments in 2014.

atmosphere or received ashore without measurement, which is usually the case. For example, upon loading such a cargo to a ship, the ship's tanks will first be filled with vapour evaporated from the cargo. This will then be displaced from the tanks by the incoming liquid cargo and will usually be ventilated to the atmosphere, representing a physical loss. (Some progress is being made in developing methodologies for the capture of such vapours and returning them to shore. They still represent a loss against the cargo quantity, however.)

Evaporation continues to occur on voyage, and diurnal temperature variations will also promote such losses. Further evaporation of the cargo will occur into the increasing ullage space on the vessel during discharge. These vapours will be left on board after the discharge, thus resulting in a further physical loss.

It is thought that such evaporation losses can routinely be as much as 0.05% during loading, 0.05% on voyage and 0.05% during discharge, suggesting that such losses can make up a major part of the inevitable losses. These losses can be increased beyond the inevitable if, for example, the pressure/vacuum (PV) relief valves on the ship's tanks are in poor condition or improperly operated.

Cargo remaining in tanks – residues: Tank emptying operations will inevitably leave some product behind, lying on the tank floor (ROB – see below) or clinging to the tank walls and internal structure; this is referred to as 'clingage'. While such losses arise with all cargoes, the residual quantity will be greater as the viscosity of the cargo increases, and can also be significant for cargoes that require heating (some crude oils, vacuum gas oil and fuel oil) to prevent/minimise waxing and solidification, especially if the heating is improperly performed. In addition, errors in tank discharging procedure or equipment failure can lead to excessive losses as residues.

Measurement of these residues is problematic. As an example, consider a ship's tanks. These are almost invariably filled and drained for every cargo, whereas shore tanks are not. Measurement of residual cargo present on the tank bottom before loading (known as OBQ – on board quantity) and after discharge (known as ROB – remaining on board) is required to monitor the ship's performance and cargo losses. This is done by dipping, ie, lowering a rod to the tank bottom and measuring the length immersed in oil/residues. Modern tankers typically have only a single 'dipping point', which will likely not be at the extreme aft end of each tank (where the residual product depth should be greatest due to the stern trim of the empty vessel). Gauging from only a single point means that no assessment can be made of variation in residual material depth across the tank. Furthermore, the decision on the calculation procedure to apply depends upon a subjective assessment of whether the material clinging to the sounding bob is liquid (in which case a standardised 'wedge' formula will be used) or non-liquid (in which case the vessel calibration tables will be used). Finally, in numerous instances the equipment used means that the surveyors cannot verify that the bob has properly reached the tank bottom due to lack of overall depth indication. Typically this is caused by lowering the bob on a rope rather than a calibrated measurement tape. In short, unmeasured ROB can be a cause of unexplained shortage.

The question of clingage is a further complication as there is no mechanism for its measurement. Traditionally, it has been estimated by relating it to the measured ROB, applying a multiplier to the former. Clingage is now, except in the circumstance of waxing (as discussed below), less of a factor than it was in the past. The tanks of modern ships have a clean internal structure, ie, all or most of the supporting framing is outside the tank, leaving only easy draining vertical surfaces inside. This clean structure also eases ROB problems. Further, with regard to crude oil cargoes, the adoption of crude oil washing of the tanks during discharge significantly reduces both clingage and ROB, except with certain waxy grades. This is discussed in the next section.

Notwithstanding the difficulties in measurement, the presence of either detected or undetected residual cargo in a tank will not necessarily lead to losses, provided that the tanks concerned are in a similar condition following their discharge as they were prior to their loading. Even if all operations are well performed, however, residue losses can be significant if the tank(s) concerned had been clean and dry prior to receiving the subject material, and were thus 'wetted' by the incoming cargo. A good example of such losses would be for crude &/or fuel oil vessel loading immediately after a dry-docking (in preparation for which a comprehensive tank washing operation would have been undertaken to remove all cargo residues).

Waxing and precipitation of solids: Some crude oils and products (vacuum gas oil, some fuel oils) are prone to precipitate wax and/or other solids as their temperature reduces.

The Wax Appearance Temperature (WAT) of an oil is the temperature at which waxy solids form in the liquid when cooled. Test method IP389 can be used to determine the WAT of liquids that are both transparent (such as gasoil) and opaque (such as crude oils). If a cargo is cooled to a temperature at or below its WAT, then wax will precipitate from the cargo. These waxes cannot easily be removed from cargo tanks as they have high pour points (the lowest temperature at which the oil will continue to flow under the influence of gravity when tested under prescribed conditions) – these are typically higher than the maximum heating capability of a vessel. Crude oil washing may not be effective once waxes have formed, although heating of the crude oil washing medium can aid the removal process.

Certain typical quality details and recommendations for the carriage and discharge of a crude oil are contained in EI booklet HM40 (*Guidelines for the Crude Oil Washing of Ship's Tanks and the Heating of Crude Oil Being Transported by Sea*).

As an example, the typical quality for Kirkuk Crude listed in HM40 is (among other things) as follows:

Wax content	: 5.06%
Pour point	: -45°C to -15°C
Cloud point	: 16°C to 19°C

The characteristics of each shipment may vary. In the absence of known quality data for a cargo, good practice for the carriage of a cargo would be to follow the HM40 recommendations carefully with regard to heating and crude oil washing.

HM40 specifies "No heat" for load, carriage and discharge of Kirkuk Crude. The 'remarks' column, however, provides as follows:

This crude is affected by ambient conditions, especially sea temperature. Heating during winter months will be needed. Heating to 25°C and bottom washing may be required.

These instructions are somewhat confusing, in that the remarks suggests heating is necessary (during winter months), despite the "no heat" recommendations listed in the load, discharge and carriage temperatures columns. The official guidance alone may not, therefore, be sufficient for a prudent charterer and master/shipowner to consider carefully whether cargo heating and/or bottom washing will be required for any particular crude oil. Disputes have arisen as to which of the preceding parties is responsible for the determination and adoption of the appropriate discharging conditions.

Other cargoes, notably fuel oils, can precipitate solids known as asphaltenes, especially when incompatible grades are mixed. These solids can form a hard compacted layer at the bottom of the storage tank or ship's tanks which is likely to be extremely difficult to remove. Indeed, it is not unusual for such material to have to be manually dug out of the tanks. Claims for shortage and for the cost of removal can therefore arise.

Misappropriation: Misappropriation (ie, theft) of cargo can arise at any stage of a shipment. The authors have investigated numerous cases of alleged misappropriation (both substantiated and unsubstantiated) of crude oil, fuel oil, gasoil, kerosene, and gasoline from multiple sources, including shore terminals, tanks, pipelines, rail cars and ships.

On a number of occasions in which misappropriation has occurred, the material has simply been removed, with no attempt to hide the fact. The theft may then become apparent when the oil is next properly measured.

On other occasions, however, elaborate steps have been taken to disguise the occurrence of a theft. A typical technique used in an attempt to mask oil removal is to pump afterwards the same volume (as the volume of the oil removed) of water back into the tanks concerned and, if subsequently detected, claim that the water "must have been part of the original cargo". In such cases, characterisation analysis of the water can often determine the likely origin of the water (if seawater, the technique can sometimes determine the geographical area from which the water originated). If the origin of the water is identified, it can greatly facilitate the investigation into the source of the water ingress and thus help identify the party responsible. For this reason, it is important that all load and discharge port samples are retained whenever a potential claim arises, even where it is a claim for shortage.

Misappropriation on ships: For seaborne shipment, removal of part of cargoes such as fuel oil or gasoil for use as ship's fuel has arisen in the past. In theory, such theft can be detected simply by the cargo inspectors undertaking a bunker survey. In the authors' experience, however, cargo inspectors seldom properly perform or even witness such measurements, often simply relying instead upon details provided to them, usually by the vessel's chief engineer. Clearly, a chief engineer engaged in the practice of cargo misappropriation is unlikely to provide reliable measurements! Modern chemical fingerprinting techniques, however, such as those used to identify the origin of seaborne oil pollution, can sometimes identify the presence of a particular oil in bunker samples, so that analysis of properly drawn bunker samples can assist in the investigation of such losses. In the event of a shortage arising, such samples should be drawn and retained for analysis, even if only to eliminate this unpleasant possible cause from consideration.

Misappropriation on shore: The authors have also investigated a number of very large claims arising from term contracts Some of these contracts relate (for example) to the supply of crude oil to a refinery in exchange for refined products. Often in such agreements, 'warrants' or 'holding certificates' are required to be issued when the oil is initially delivered. However, as the oil is typically in common storage with other oil and is necessarily used to produce the required products, the certificates themselves often show only the received oil and contain no assessment of the overall current physical stock at any time. Thus, there is no periodic verification of physical stock against book stock.

Under such arrangements, the refining losses may be greater than had been predicted, leading to a lower than expected yield of refined products, or oil may be fraudulently exported from the refinery without authorisation. In either case, a situation will eventually arise in which insufficient material remains in the terminal to allow the stock holder to meet its obligations to the supplier. The shortfall can be hidden by processing subsequent incoming cargoes but leaving them short, in a form of 'Ponzi scheme', until eventually even this process cannot hide the existence of a shortfall. Only at this stage is a stock check performed that is sophisticated enough to determine what has happened.

Misappropriation during multimodal transport: A similar situation sometimes arises when a trader moves large quantities of oil from place to place over a period of time and involving multiple transport and storage arrangements. Oil leaves point A and is transported to point B by a route perhaps involving trucks, trains, ships, pipelines and tank storage. This journey can take a significant period of time. Oil is measured when it enters the system and when it leaves. Inevitably, the latter figure is a lot less than the former as the operation proceeds. It is presumed that the difference is 'inventory' within the transit system. Only when the operation is complete are losses from the system, including those caused by misappropriation, revealed.

2.2 Causes of apparent losses

Apparent losses can arise due to errors and/or inconsistencies in the measurement or calculation of cargo quantities. Despite being only apparent, rather than real, they can nonetheless be the cause of claims. Different types of apparent loss arise, notably measurement tolerance, bias, error and inconsistency. These are discussed below.

(a) Measurement tolerance

There are numerous different means or procedures for the measurement of oil, often determined by the point at which they are measured – for example, in shore tanks, ship's tanks, road trucks and rail cars, or while flowing through pipelines. Each procedure is in some way dependent on the determination of the density, temperature and volume of the oil at that temperature and each determination will be subject to measurement precision or tolerance. Thus apparent shortages between bill of lading and outturn measurements may arise simply due to the tolerance of the measurement sources involved. It will be realised that a 'shortage' arising out of measurement tolerance is neither real nor an error. Common assessments of the extent of such tolerances are that repeat measurement of the same quantity of oil in a shore tank might show a tolerance of less than ±0.1%, measurement of the same cargo on a ship at load, and discharge port ± 0.25%, and that the effect of measurement tolerance on overall custody transfer might be around ± 0.3% (the same as the EI study standard deviation).

(b) ***Bias***

Further, some measurements suffer from inherent 'bias', which leads to a regular over- or understatement of quantity – for example, tank calibration error. Cargo inspectors routinely monitor bias in a ship's measurements by determining the vessel experience factor (VEF) – a correction factor derived by comparison of ship and shore figures over a series of loadings. Thus, if the VEF is properly calculated, any bias inherent in the vessel's measurements can be corrected. Furthermore, a VEF-corrected ship's figure can be used as a good comparison for shore delivery and receipt figures in order to assess whether there is any bias and/or error in the reported results. Bias can give rise to claims for shortage, even though the loss is only apparent.

(c) ***Errors***

The contents of both ship and shore tanks can be stratified, containing oil with layers of differing temperature and/or density. Such temperature/density layering can present difficulties for accurate sampling (and hence density determination) and/or measurement of temperature.

Oil measurement procedures specify that tank contents should be assessed for layering by individual measurement of both temperature and density at upper, middle and lower (UML) levels. If any significant differences between the layer results are observed, then samples/measurements should be undertaken at 'multiple' levels in order to obtain representative results.

Cargo inspectors, however, may take only upper, middle and lower measurements/samples and then average the results so obtained. This can give rise to significant error. For example, if a shore tank contained (say) 4,000 cubic metres (m^3) of 'cold' fuel oil at 20°C, and then received 50,000m^3 of 'hot' fuel oil at 60°C, it should then contain 54,000m^3 with an average temperature of about 57°C. It is, however, likely that the hotter, (and thus lower density) material would sit 'on top of' the colder and more dense material, thus leaving the tank contents layered. If it is assumed that no mixing whatsoever occurred, a cargo surveyor taking 'accurate' measurements of UML levels would obtain temperatures only in the 'hot' fuel oil layer and obtain a temperature of about 60°C, thus overstating temperature. A minor

Figure 1: A shore tank with temperature stratification

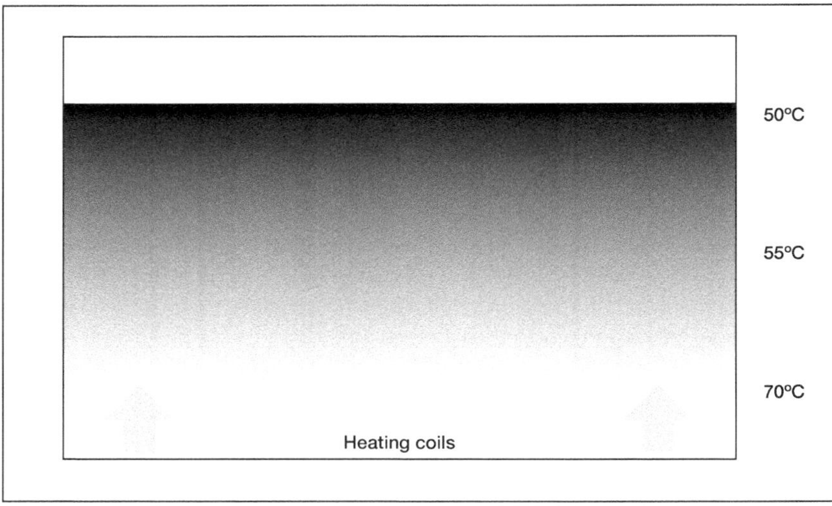

error in depth measurement (the lower measurement should be taken 1/6th of the depth from the bottom, which would be just above the top of the 'cold' layer 1/10th from the bottom) could, however, result in the measurement being taken wholly or partly in the cold layer, which in the extreme could result in a measurement of only 20°C. An average of UML measurements obtained in such circumstances would equate to 50°C and thus considerably understate temperature.

Each degree of temperature error causes a 0.1% error in measured volume.

(d) ***Floating roof tanks***

Some shore oil storage tanks have floating roofs – ie, they float on the oil and reduce evaporative losses as well as increase safety (by reducing the risk of fire and explosion). These roofs can weigh hundreds of tons in the case of larger tanks. When the tank is empty, the roof is supported on a series of 'legs' that protrude beneath the roof and rest on the tank floor. Once the tank is filled to a depth in excess of the length of the legs, however, such roofs will displace their own weight of the product (the Archimedes Principle) and level gauging will show an apparent increase in the product depth not corresponding to any increase in the quantity of oil in the tank. Thus, quantities determined from tank level gauging must be adjusted for the full floating roof weight when the roof is freely floating. This adjustment is determined by calculating the volume of oil at the tank temperature and density that would be displaced by the roof weight.

Significant error can arise if the roof is in the 'critical zone' during tank measurement. This is the zone based on tank liquid levels between which the roof is partially floating and partially supported by its legs. Thus, only part of the roof weight is being displaced in the liquid. While accurate tank calibration in such conditions is theoretically possible, the effect of factors such as warping of the tank floor, debris (rust and sludge/waxes) on the tank floor or warping of the roof cannot be estimated. The calibration tables are therefore generally considered to be unreliable within the critical zone (and typically contain a caution to that effect). Measurement with the roof in this state is likely to introduce an error and so should be avoided.

(e) ***Pipelines***

Loading/discharging terminal tanks can be situated significant distances from a vessel loading or discharging a cargo. In such circumstances, it is inevitable that there will be a pipeline of substantial volume connecting the two. It is important that the pipeline contains the same quantity at the time of measurement before and after the movement, otherwise substantial measurement errors can arise unless the difference can be properly accounted for.

Many pipelines are ostensibly left 'full' after each movement. If, however, ship and/or shore tanks have been emptied during the movement, then air and/or gas from the stripping process can displace some of the pipeline liquid contents. Self-evidently, air/gases in a pipeline that are assumed to be oil can give rise to measurement errors.

Such circumstances would not only affect the subject cargo transfer; there could also be an effect on the following movement, which might (for example) start with a part-full (slack) pipeline and complete with a full pipeline. Some of the cargo transferred in that case would simply have been used to refill the pipeline and would not reach its intended tank, leading to apparent losses and thus claims.

For the above reason, it is prudent for the cargo inspectors to undertake a 'pipeline displacement check' at the commencement of a loading or discharge operation. This is accomplished by the vessel loading or discharging the approximate pipeline volume to or from one or two tanks on board to or from a single shore tank. Comparison of the delivered and received quantities can then be used to determine how much of the cargo has been used to 'refill' the pipeline, allowing the calculation of loaded or discharged quantities to be adjusted accordingly.

(f) ***Loading from/discharge to RTVs & RTCs***

Some operations can involve vessels loading from or discharging to road tankers (RTVs) or rail tank cars (RTCs). For example, many shipments to/from former Soviet Union (FSU) countries have involved RTCs. The quantity delivered to/received by an RTC is typically determined by weighbridge measurement or by measuring the volume from the tank sounding by reference to calibration tables. There are, however, a number of reasons why such measurements can be unreliable. For this reason, the contract between the sellers and buyers of the oil can often specify that bills of lading/outturn etc. are to be determined solely on the basis of the vessel's VEF corrected load/discharge figures, thus avoiding claims for apparent losses.

(g) ***Ship-to-ship (STS) transfers***

STS transfers of oil are common nowadays. This often involves one vessel acting as a 'mother' vessel, receiving oil from a number of 'shuttle' tankers and then undertaking another transfer to an export vessel.

Cargo quantities for STS transfers are generally obtained using measurements from one vessel or a mathematical average of both of the vessels involved. On both vessels, obtaining accurate measurements in an offshore location can be difficult as a result of rolling and pitching of the vessel in a seaway. This is particularly the case where measurements are determined using modern closed tank gauging and sampling systems.

Furthermore, there is often a significant difference in the size of the vessels involved, with the mother vessel often being considerably larger than the delivering shuttle tankers and, possibly, export tankers also. In such circumstances, a measurement tolerance of (say) 0.5% (to reflect the imprecision of measurement in an offshore location) on a 100KT mother vessel could result in up to 1% error in the measurement of a 50KT export cargo. Furthermore, the transfer of only part of the cargo from a mother vessel would also introduce potential problems with regard to the VEF (which is only deemed to be applicable for full cargoes), and the condition of the vessel's pipelines before/after each transfer.

(h) ***FPSOs***

There are large numbers of floating production and storage offshore (FPSO) vessels, which receive, process, store and then directly transfer oil to export vessels. Delivered quantities from FPSOs can be determined by measurement of the cargo tanks before and after the transfer. The offshore environments means that they can be subject to similar errors as those discussed above for STS operations. Furthermore, VEF data cannot be determined and is thus seldom applied to FPSO measurements.

Determination of the water content of a delivered cargo is often problematical. Sampling from the FPSO is typically via closed sampling hatches, which make accurate sampling difficult. Furthermore, many FPSOs only have rudimentary 'laboratories' on board used to determine the water content of the samples drawn. Finally, it is not uncommon for the floating hose string used to connect the FPSO to the export vessel to be left full of water before/after each transfer. This water is sometimes not properly accounted for.

The authors have investigated a number of losses in which the water content of an oil determined at the discharge port is considerably greater than that reported by the FPSO at the time of loading, thus resulting in significant losses. The investigations have sometimes suggested that the FPSO water content reading was in error. Retaining load and discharge port samples for analysis during investigation of claims is therefore of great importance.

(i) ***Net quantity determination***

The net quantity of an oil is calculated by deduction of free and entrained water content and sediment from the total calculated volume of an oil cargo. Errors can therefore arise if the free and/or suspended water content of an oil is not properly determined at load and/or discharge port.

For accurate measurement of the depth of a free water layer beneath a bulk of oil, it is necessary to allow sufficient time for suspended water droplets in the oil to settle to the bottom of the tank concerned. The time required can be considerable, easily a number of days in the case of large ship or shore tanks. Commercial pressures are, however, such that sufficient time may not have been made available.

Free water measurement by manual sounding can be problematical in both ship and shore tanks. It is achieved using a water-sensitive detection system on the lower end of a measurement tape. The nature of the product may make identification of the interface or 'cut' between water and oil difficult. The bottom of the tank being measured may contain wax, sludge and/or rust scale. Finally, tank calibration in both ship and shore tanks lacks accuracy in the bottom of tanks.

In some terminals, free water dips are not in fact permitted. Instead, surveyors are required to witness the opening of water drains and check visually that there is no free water present or allow any that is present to be drained away and the quantity determined by measuring lost volume. The authors would comment in this regard that oil/water emulsions in fuel oils and/or some crude oils can be mistakenly identified as oil or as water – this may result in substantial quantities of water remaining in the tanks undetected or wrongly deducted from oil quantities.

Sampling of shore tanks is typically undertaken at upper, middle and lower levels only. Such samples, even if properly drawn, may not represent the lowest part of the tanks, which would likely contain the wettest material. Thus, the typical sampling undertaken means that water content determined solely from shore tank sources can sometimes be understated.

Similarly, measurements taken immediately after the loading of a vessel can understate the quantity of free water subsequently found in a cargo. In these circumstances, the quantity of free water detected after loading is considerably below that detected upon the vessel's eventual arrival at the discharge port.

This in itself would not be a problem in the determination of overall water content in a vessel's cargo, provided that accurate samples had been drawn from the vessel's tanks both after loading and prior to discharge. In these circumstances, the total of suspended and free water should remain constant between load port and discharge port. Limitations caused by (for example) closed sampling systems mean, however, that it is difficult and time-consuming to draw representative samples from numerous vessels' tanks. Thus, it is not uncommon for the apparent total of free and suspended water to show a considerable increase upon the vessel's arrival at the discharge port, even if no additional water had entered the tanks during the voyage.

(j) ***Inconsistency***

As previously discussed, oil calculations can be undertaken in a number of different ways. Most of the calculation procedures utilise measurement of oil volume, temperature and density in order to calculate standard volume and the weight of an oil shipment. On occasions, however, the starting point of measurement may be weight (for example, determined by measuring road tankers on a weighbridge). Thus, any error in temperature and/or density measurement of an oil will cause a discrepancy in both weight and volume-based measurements.

Furthermore, measurement systems can be fundamentally inconsistent. For example loading figures (bill of lading etc.) for cargoes shipped from FSU countries to non-FSU countries will likely be determined in accordance with the Russian

Gosstandart (GOST) system, while the discharge port figure (including outturn etc.) will be determined in accordance with the American Society for Testing and Materials (ASTM)/International Organization for Standardization (ISO) system. This will result in discrepancies. Also, loading figures for cargoes shipped from the Middle East will likely be determined in accordance with ASTM Table 6 or 54 (issued in 1952), while discharge figures will be determined per Table 6A or 54A (issued in 1980, revised in 2004). As the graph in Figure 2 below (which shows the percentage difference between calculations using tables 54 and 54A) illustrates, this can result in a significant discrepancy.

Figure 2: Table 54 versus Table 54A

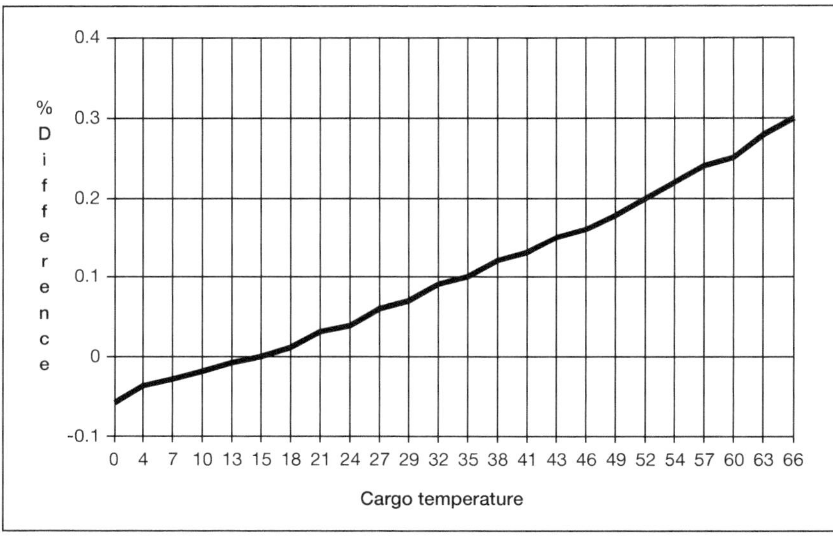

Cargoes loaded from the Middle East are almost invariably well above 15°C at loading, due to the ambient temperature. Figure 2 above shows that the apparent loss arising from an hypothetical shipment of crude at a loading temperature of (say) 35°C would result in an apparent loss of about 0.1% upon discharge at a port where Table 54A is used.

Quantity discrepancies can also arise between weights expressed in vacuum (that is, mass), and weights expressed in air. For most oils, a bill of lading expressed in vacuum compared with an outturn expressed in air would result in an apparent loss of around 0.1% to 0.16%.

2.3 Summary

It is a common principle of the insurance industry that claims should be considered on the basis of comparison of like with like. This principle is fully applicable to the consideration and comparison of oil quantities. For such comparisons to be meaningful, the quantities must be calculated by the same methodology, using the same correction tables and expressed in the same units. Except in relation to quantities determined by weighing, for example, by weighing road tankers, comparison of quantities in volume (rather than weight or mass) will be more reliable.

3. Quality determination – quality claims

Determination of the quality of a parcel of oil is dependent on the proper implementation of two key steps. First, a fully representative sample of the oil must be drawn, handled and stored by means compatible with the nature of the oil and the analytical tests to be performed. Secondly, the samples must be properly analysed in accordance with the appropriate and specified test methods.

Disputes over quality arise when a receiver of goods claims them to be of unsatisfactory quality, in contradiction to the shipper's or load port inspector's reports showing them to be in good order. In such circumstances, the term 'quality dispute' is often substituted by 'contamination claim'. This is, however, incorrect and misleading. A quality dispute may or may not have arisen as a consequence of contamination. There are a number of other circumstances that can result in a real or apparent change in oil quality, absent any contamination.

Generally speaking, quality disputes may arise for one or more of seven reasons:

(a) the cargo was always off-specification, ie, prior to loading, despite reports to the contrary;
(b) the cargo is in fact on-specification, as received, despite reports to the contrary;
(c) different and incompatible test methods are being used;
(d) measurement tolerances;
(e) the cargo suffers from inherent instability;
(f) the cargo is physically damaged; and
(g) the cargo has been contaminated.

Scenarios (a)–(e) may be thought of as 'apparent' contaminations. Scenario (f) relates to physical damage to the oil (for example, as a result of overheating) but does not represent a contamination incident. It is only scenario (g) that involves a 'real' contamination.

Establishing which of these scenarios has (or have) occurred is of critical importance when investigating contractual quality disputes. The success of any such investigation is intrinsically linked to the quality and reliability of the samples presented for testing. Unfortunately, taking representative samples of oil cargoes is not without difficulty and some of the challenges that parties involved in such activities are faced with are discussed below.

3.1 Sampling

Sampling is of fundamental importance. Without a representative sample, no amount of analysis (no matter how sophisticated or expensive) can determine the quality of the parcel.

The challenges of obtaining a representative sample are substantial. Taking an extreme example, a VLCC cargo of crude oil may have a volume of around

300,000m^3. The sample drawn to represent it might amount to five litres. Thus the sample taken would amount to less than 0.02 parts per million of the cargo. Further, only one millilitre of this sample may be used to determine, for example, the density. Thus the amount analysed corresponds to around 0.003 parts per billion of the cargo and, in order to obtain reliable and correct results, the amount analysed must be representative. Given that density is ultimately used to calculate both volume and weight (and therefore value) of the cargo, the importance of accurate and reliable results cannot be underestimated.

To ensure samples are representative, they should be taken in accordance with methods set out in published international standards. A number of different sampling standards are issued by industry bodies such as the ISO, the EI or the ASTM. For example, ASTM D4057-12 details various procedures for manual sampling of oils.[5] The purpose of these documents is to standardise the conditions for obtaining samples of the products in question.

Sampling is particularly difficult where the oil that is to be sampled is not homogenous – for example, crude oil (perhaps having increasing water content with depth), gasoline/naphtha (density layering) and any oil that has suffered some form of contamination or quality deterioration. If drawn in line with the procedures outlined in the standards, however, the samples should allow the degree of non-homogeneity to be assessed and the quality of the material to be estimated.

There are many differing procedures that can be used for sampling. Traditional manual sampling is the most common. At its most straightforward, manual sampling involves lowering a can to particular depths in the tank, at which a cork is removed to allow the can to fill with the material present at that depth.

(a) *Shore side sampling*

Typically, samples from shore tanks before/after shipment are drawn from upper, middle and lower layers. These samples may be tested separately to assess homogeneity or may be combined to make a composite sample.

If, as is commonly the case, a composite sample is made by combining upper middle and lower level samples, it may not be representative if the oil is layered. To assess this, standard sampling procedures require it to be determined first, before blending the composite sample, by comparing the densities of the upper, middle and lower samples drawn. If these are the same within the repeatability of the test method, then the tank's contents can be considered homogeneous and the level samples combined, with the composite produced then being considered representative of the whole. If the densities are different, then the tank is considered layered and should be sampled at each metre of oil depth, with all of these samples being combined to produce a composite.

These difficulties are largely avoided (although to some extent replaced by others) by drawing the samples with auto-sampling equipment placed in the shoreline through which the cargo is to be loaded or discharged. Such samplers are commonly fitted into crude oil shore lines, both at load and discharge ports. Auto-

5 *Standard Practice for Manual Sampling of Petroleum and Petroleum Products.*

Figure 3: Sampling from shore tanks

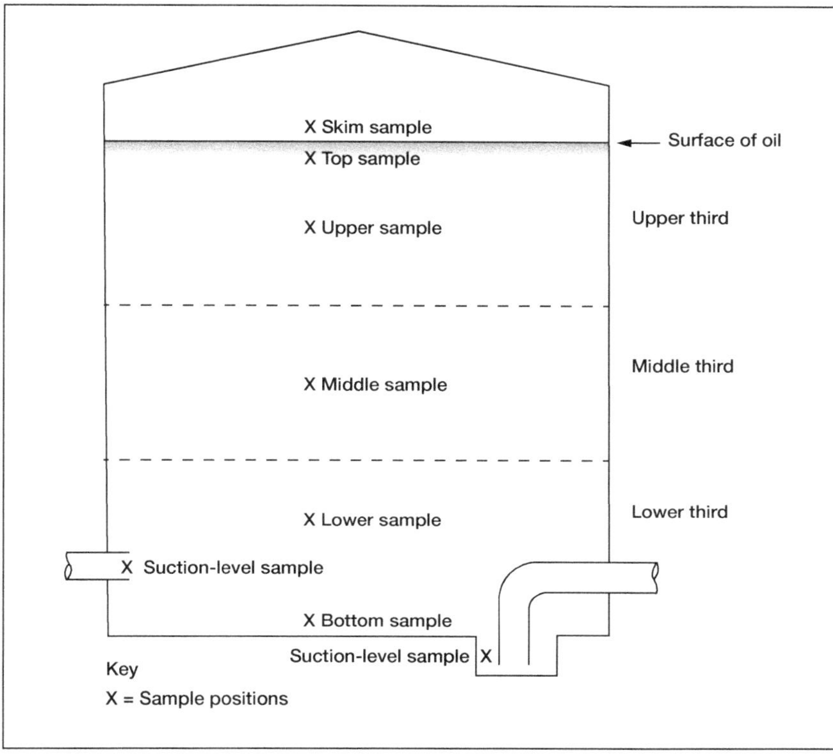

sampler design is a sophisticated process. Modern designs draw samples on a flow proportional basis: that is, they are made so that a small aliquot is drawn from the pipeline for the flow of a given volume of cargo. It is important that the sampler is positioned in a section of the pipeline where the oil is homogeneous across the pipeline cross section and mixing systems can be fitted to achieve this. Turbulent flow at the sampler location is desirable.

For an independent inspector monitoring a loading or discharge, auto-samplers are, to an extent, 'black boxes' over which he has no control. Prior to the start of an operation, however, they are programmed with criteria such as the expected volume of the cargo to be sampled and the desired size of the sample to be drawn. This determines the number of aliquots to be taken, which should be at least 10,000. Confirming at the end of the operation that the appropriate number of aliquots have been drawn and the sample volume obtained corresponds with aliquot number x aliquot volume are standard checks that can be made on the sampler's performance. Manual samples from the shore tanks and ship taken at the different stages in the operation are useful in the investigation of unexpected results. Ideally, these should be taken and retained.

(b) ***Ship side sampling***

Nowadays, most seagoing oil and chemical tankers only permit 'closed' sampling to be undertaken in routine situations. The equipment used typically consists of a zone[6] sampler that is lowered into the cargo tank through vapour-locks.

In theory, a closed system zone sampler can produce representative samples. In practice, however, there are a number of issues with samples drawn from such devices. The samples may suggest that sound cargoes are off-specification, or that off-specification cargoes meet specification. Therefore, in the event of a potential quality concern following closed sampling, it is important to attempt to obtain traditional 'open-hatch' samples before any decision is taken on the fate of the cargoes concerned. Many terminals and/or vessels will not, however, permit open hatch sampling, citing safety/environmental grounds.

Auto-samplers are rarely used for cargo sampling on ships, although simple versions are frequently used to sample bunker fuel loadings.

3.2 Analysis

Once representative samples of an oil have been obtained, accurate analysis must be performed using appropriate equipment and methodology.

All petroleum specifications should, as well as setting out the relevant quality parameter limits, define the test methods (often published by ISO, EI or ASTM) to be used for their determination. This is important, as there are likely to be a number of different methods that could be applied to a particular parameter (the determination of sulphur content being a good example) and these will not necessarily produce consistent and comparable results. Testing performed using different test methods at load and discharge port is a common cause of quality disputes.

It should be noted here that it is not always sufficient to assess the quality of an oil solely in relation to the criteria of quality set out and the values defined in the specification. Often by law, it must also be fit for its intended purpose. For example, cyanide content is not specified for orange juice but the inclusion of cyanide at a significant level in the juice would certainly be important! Some petroleum product specifications now address such issues, as well as defining specific limits for a list of parameters.

An example of this, and arguably the most common quality specification for marine fuels (that is, those petroleum fuels for use in marine diesel engines and boilers) is ISO 8217:2012 *(Petroleum products – Fuels (class f) – Specifications of marine fuels)*. This standard includes tables of testing requirements and states that the fuel specified under the standard "... shall conform to the characteristics and limits given ..." in the tables "... when tested in accordance with the methods specified." ISO 8217, however, also sets out a number of other requirements in addition to the fuel needing to confirm with the listed test requirement limits. These further requirements include a statement[7] that the fuel "... shall be free from any material

6 Defined as a sample taken at that part of the liquid column which is trapped within the whole height of a sampler when it is sealed at a single spot location within a tank, after having been fully flushed as it was lowered to that position.

7 Taken from the 5th Edition of International Standard ISO 8217:2012(E).

that renders the fuel unacceptable for use in marine applications". It is simply not possible to define and prescribe limits for all the materials that may render a fuel unacceptable for use in marine applications.

In the same way, petroleum products traded under contracts that do not have such explicit clauses may be contaminated by materials that do not form part of the specification, but may lead the receiver to question whether the fuel can be safely used for its intended purpose. It is typically the case that demonstrating whether or not the fuel is fit for purpose is complex, requiring a detailed and thorough analytical assessment of the quality of the fuel above and beyond that which would normally be done as part of any routine testing. Obtaining sound legal advice as to the contractual position is also always well advised!

3.3 Types of quality claim

(a) *'Apparent contaminations'*

Earlier in this section, the authors identified a number of situations in which one or more of the parties to a trade may suspect that there was a contamination incident leading to a quality claim, but where the suspected 'contamination' was in fact attributable to some other explanation. Some examples are discussed below.

Sampling issues: Very often, quality disputes are attributable to sampling issues which may, for example, be caused by the use of closed system sampling devices described previously. For example, closed system samplers are often fitted with a 'wiping' device to remove liquid from the lowering and lifting tape as the sampler is extracted. This causes liquid to run down the tape into the top of the sampler. This has been known to lead to extraneous contamination in samples, leading to suspicions of contamination in the bulk cargo where, in fact, drawing of open hatch samples has subsequently proved that there was no contamination.

When drawing samples of a cargo for the purpose of assessing quality, it is important to remember that certain quality parameters require specific sample handling that may not be compatible with the requirements of other test methods and may therefore require separate samples to be drawn. This is of particular importance when investigating reported quality claims – for example, the use of plastic bottles with non-airtight caps would not be recommended if investigating problems with the flash point[8] of a distillate cargo, such as jet fuel, due to the potential loss of volatile components from the sample that may give rise to misleading results. If, however, it is suspected that the cargo has been contaminated with trace levels of metallic species, plastic bottles are preferred to ensure that the samples have not inadvertently been contaminated with metals emanating from a metallic sample container.

Testing issues: The importance of using the same standard test method to determine a particular quality parameter at different stages in the cargo movement has been

---

8 Flash point is a parameter specified for safety reasons which, simplistically, measures the temperature at which a mixture of air and the vapours produced by a sample is flammable.

discussed previously. In addition to the fact that the use of different test methods may produce results that are not comparable with each other, standard test methods usually include a precision statement within the method to take into account the expected variability or imprecision associated with the test method.

The measurement tolerance (repeatability and reproducibility) inherent in most analyses means that some variation in test results can be expected even if the samples are tested in strict accordance with the procedure detailed in the standard test method. This could mean that a quality parameter found to meet specification at the load port is subsequently reported to be off-specification when tested at the discharge port, despite no actual change in the quality of the material.

Reference to a 'Precision Statement' within a test method allows test results produced by the same laboratory, or by different laboratories, to be compared, when samples of nominally identical test material have been correctly tested in accordance with the procedure outlined in the method. Using the precision information in the test method, it is possible to identify whether differences in test methods are likely due either to the expected variation in results or to some other factor, such as contamination.

A number of sets of guidelines have been produced by those bodies responsible for preparing and issuing test methods that provide instructions to parties disputing test method results. Common practice within the industry involves retesting samples under conditions intended to minimise the expected imprecision associated with the test method. This is often done by organising a 'joint analysis' of samples from all stages in the cargo's movement, in which the interested parties each send representatives to witness the testing performed in accordance with a standard test method or methods at a suitably equipped and mutually agreeable laboratory. Such 'joint analyses' should resolve disputes regarding the actual test results and, if the relevant samples are made available, should allow the representatives in attendance to advise their principals whether any suspected change in quality is attributable to the precision of the method or whether there has been a 'real' change in quality.

(b) 'Real' quality changes

This section discusses some of the factors that can give rise to an actual change in quality of a material. Such quality changes can arise as a result of natural deterioration in quality (known as inherent vice), as a result of physical damage to the cargo and/or by contamination with some other material.

Inherent vice: Specified quality parameters of an unstable material can change over a period of time without any specific, unusual, external influence. In such circumstances, a material can be considered to have an inherent vice.

Examples of changes in quality arising from an inherent vice include an inevitable colour change in some gasoils, arising over time and resulting from the instability of one of the blend components used to produce the final gasoil blend, or the inevitable increase in gum content of gasoline caused, perhaps, by reaction with oxygen. Both of these examples can result in a cargo that was initially on-specification being correctly determined as off-specification at the discharge port, even though there has been no contamination or mishandling at any stage in the cargo movement.

Physical damage/mishandling: Certain cargoes may be sensitive to the particular conditions under which they are carried and changes in quality may be promoted by exposure to, for example, too little or too much heat. In such instances, resultant changes in quality to the cargo may render it off-specification, without any contamination having taken place. Typically, a detailed review of the carriage instructions issued to and the voyage data recorded by the shipper will be required to ascertain whether or not the shipper has complied with the instructions issued.

Contamination: If a material is inadvertently mixed with another material of differing quality, a contamination is said to have arisen. Many oil products and petrochemicals are shipped on vessels that are not dedicated for that particular cargo, so that there is potential for contamination from other cargoes being carried, either concurrently or as previous cargoes. Some common examples of this follow.

Previous cargo residues: Residues of previous cargoes can be present in a vessel's tanks and/or lines prior to loading, due to inadequate cleaning. This can be difficult to detect (both by the vessel's crew undertaking the cleaning and the surveyor performing the tank cleanliness inspection), due to the requirement for the tanks to be presented for loading under an atmosphere of inert gas, thus restricting the inspection to closed gauging. Contamination from previous cargoes is usually more serious (and more readily detected) for highly specified products such as aviation fuel, rather than crude oil.

Cargo cross-contamination: Many vessels are expected to load, carry and discharge multiple cargo grades 'within vessel's natural segregation' (WVNS). This is normally expected to mean that the pipelines used are separated by at least two valves. In order to maintain such segregation, however, it is often necessary to load and/or discharge cargoes individually, whereas commercial pressures often require cargoes to be handled concurrently.

This can often leave just a single valve between grades, and any leaks and/or errors by the crew in setting the valves can lead to cross-contamination.

Vapour phase contamination: The requirement for a vessel to segregate the cargoes carried properly also applies in the vapour phase. Thus, there should be isolation in the inert gas (IG) main between different cargo grades.

This is particularly the case where volatile, high vapour pressure cargoes such as gasoline are carried concurrently with low vapour pressure cargoes such as gasoil. If isolation is not maintained, the difference between the vapour pressures of the cargoes means that volatile, low flash point components from the gasoline will pass through the IG line and be absorbed into the gasoil, severely affecting the flash point.

4. Blending

While the preceding section has dealt with the inadvertent mixing of oils, blending – deliberate mixing – is ubiquitous in the petroleum industry. Most crude oils are blends of oil produced from different wells in the same general field, and many export streams are blends of the production from different fields within the same region. The crudes mixed can have quite widely different qualities. The objective in making these blends is often to create a mix of sufficiently large volume that it can be marketed at one specification, as well as to use up both the most and the least desirable oils produced from individual wells. Despite this process, the quality of export streams can vary markedly, both over a long period as the quality of production from a field varies, and perhaps over a short period as particular component crudes from key fields normally included in a blend become unavailable for operational reasons.

Further, refiners themselves will often make and run through their refineries a blend of different crude oils as together they perform synergistically during the refining process, most likely by allowing each of the refinery process units to operate simultaneously at maximum efficiency – something that would not be achieved by running the crudes separately. To facilitate this process, some organisations deliberately make and market to refiners a pre-blend of crude oils from quite different fields.

With regard to marketed 'finished' petroleum products, these are almost always blends of individual 'intermediate' streams produced within the refinery. They are made from these intermediates by blending them together in a mix designed to meet the particular target specification to which the product is to be supplied.

As well as finished products being blended from intermediate refinery streams within the refinery as part of the overall process of refining, many products are blended away from refineries in tank storage terminals or on ships. Such activity is often undertaken by trading companies that have seen an opportunity to create value by mixing together lower priced components to make a product of greater value than its component parts or indeed mixing them together to make a lesser grade finished product that meets a specification without 'product quality giveaway', meaning that it is of no better quality than it need be.

Two stages involved in producing the blend in all of these cases: firstly to design it and secondly to mix together successfully the component parts that the design requires to be incorporated.

4.1 Blend design

Blend design is a multi-faceted process. The different available blend components will be available in different quantities, have different qualities in respect of many different test criteria and, importantly, be of different financial values. The objective is to design a mixture of these components that will meet the target specification with minimal quality giveaway and at minimum cost, while ensuring that all available products can be fully utilised within the overall operation. Not surprisingly, the design process is usually fully computerised, with refiners and outside blenders having sophisticated programs for the task.

Inherent in this process is consideration of the test properties of each of the blend components against the required target specification. The mix must be designed in such a way that each specification parameter is met. It must be remembered that introducing more of a component to meet one quality parameter may result in the worsening of another. Many design iterations may be needed to meet all requirements. For this reason, linear programming models are used.

Blend calculation procedures vary from parameter to parameter. For example density or specific gravity (but not API), water content, sulphur content (and generally all other parameters described as 'content') and distillation range will blend in a linear fashion: that is, the result will be expected to be the weighted mean of the component parts. Other parameters, such as viscosity, flash point, vapour pressure, octane rating and the like, do not, however, blend in this way. These generally rely on artificially developed blending indices for each component that are derived sometimes on the basis of scientific principles and sometimes from experience. In the past, such indices were presented in tables or as nomographs. Neither tables nor nomographs, however, lend themselves to ready incorporation within computer programs. Accordingly, wherever practical, the industry now tends to use the algorithmic approach to predicting the blending of non-linear qualities.

As an example, the prediction of the viscosity of a blend is normally performed by the Refutas method, which is a proven approach dependent on blending indices. It utilises the following formula to calculate blending index:

Refutas Blending Index = 33.468*(LOG(LOG(Viscosity+0.8)))+23.097

In this formula, viscosity must be entered in terms of kinematic viscosity at 50°C in cSt (centistokes). The indices are combined in weight proportion and the formula is 'reversed' to derive the blend viscosity for the combined Refutas index. This calculation procedure is known to be accurate and reliable. Its use in predicting the viscosity of a blend is very important, as the following example illustrates.

Consider the design of a blend of 380 cSt fuel oil to be made from residual oil with a viscosity of 901 cSt and light cycle oil (LCO) of viscosity four cSt. If a linear calculation is performed, the predicted design would require the blend to be formulated from 40.1 wt% residual oil and 59.9 wt% LCO. The use of blending indices in the Refutas system shows, however, that the formulation actually required is a blend of 90.8 wt% residual oil and 9.2 wt% LCO. The implications of this difference, in terms both of producing on-specification material and of economics, are substantial.

As a second example, flash point will be considered. This is the temperature at which the hydrocarbon concentration in the air above the sample in a test apparatus becomes high enough for combustion to occur when a naked flame is introduced into it. Flash point is thus related to the vapour pressure of the liquid being tested at any particular temperature. This can be readily predicted for a liquid consisting of a single chemical species, but is much harder for a mixture of unknown ultimate composition, such as a petroleum product, not least because it does not blend linearly.

Flash point is, however, a critical specification to many petroleum products, and as such it is important to be able to predict the flash point of a blend that is being formulated. In response to this task, a number of researchers have, over the years, developed methods of predicting the flash point of a blend, on the basis of the flash points of the individual components and their respective concentrations. Experience has shown that these methods can provide a reasonable prediction of the flash point of a petroleum mixture, provided that the methodology has been correctly applied and the flash points of the individual blending components are within the scope of the applicable test method.

The nature of petroleum mixture is such that none of these formulae can be derived from any fundamental relationship and they are all dependent upon numerical constants factored into equations derived empirically. For this reason, it is vital that, when using a method, the data is entered in the intended units.

For example, flash point can be reported in any one of four different temperature scales, namely Fahrenheit, Rankin, Celsius or Kelvin. A specific formula will have been developed on the basis that flash point must be entered in one specific unit, eg, degrees Fahrenheit. Similarly, concentration can be expressed in weight %, volume % or molar %, (The molar proportion of a component in a mixture represents the number of molecules of the component in the mixture, relative to the number of molecules present of all components. Rather than counting the number of molecules, molar proportions are expressed as the number of gram moles (ie, the weight of the component divided by its molecular weight) of each component in relation to the number of gram moles of all components.)

As an example, the flash point blending formula commonly used in the oil industry is based upon one published by BP, which in turn was based upon research work conducted by Butler *et al* of Imperial Oil Research (Exxon). Indices are combined in molar proportions of the constituents.

By its original design, temperatures were to be entered in degrees Rankin. The original formula has, however, been modified by the authors to allow the flash point to be entered in degrees Celsius, which is the most common way of expressing the flash point of a petroleum product. In the authors' modified Butler formula, the Flash point Blending Index (FBI) of a blend component FBI is calculated as follows:

$$FBI = e^{(8500/((T*9/5+32)+460)-11.486)}$$

The symbol T represents the flash point of the component entered in degrees Celsius. Having determined the FBI of each component, combining them in proportion to the molar concentration of each allows prediction of the flashpoint of the mix.

These calculations may seem complex but the most important point is simple. Any specific formula will have been developed for the input of data in specific units. Data entered into a formula in incorrect units will necessarily produce an erroneous result. Entry of data in the wrong units is a common error.

Calculations have to be performed in relation to each quality parameter of relevance and, as previously stated, it is likely that many iterations will be required before the most economical (or practical) on-specification blend design is obtained.[9] The importance of checking the design by making and testing a trial blend is emphasised.

9 A good guide to these processes may be found in JH Gary, GE Handwerk & MJ Kaiser, *Petroleum Refining Technology & Economics* (5th Edn, 2007).

4.2 Making the blend

Within a refinery, blend design is often conducted in line – that is, the blends are made on a continuous basis from run down (direct production from the refinery units), rather than batch by batch, which is the norm for outside blenders. In line analytical instrumentation can determine quality in real time and the blends are then made using live information. In fact, the live information can be used to change process conditions so as to vary the quality of the product if needed.

Some refiners take the process further and only produce intermediates, which they store until a cargo is required that must meet a particular specification: that is, no finished products exist in storage. These are blended in pipeline while exporting the cargo, usually through a special mixing unit, from the various stored intermediates during the loading process. A difficulty that arises with this approach is that there is no pre-loading shore tank sample (or any other type) showing the cargo to be on-specification. The first available sample of the cargo as a whole is the ship's composite after loading. If this is off-specification, then there is often uncertainty whether the fault lies with the shore blending system or the vessel for having caused contamination.

In other cases, various intermediates are kept in store in separate shore tanks and the blend made by transferring these into a further tank where the component parts are mixed. The tanks may be fitted with internal fan-like devices, known as mixers, or with small angled inlet pipes known as jet mixers or with a gas sparging (bubbling) system, usually fed with nitrogen, to promote mixing of the components. In the absence of such systems, mixing may be achieved by pumping the material round a loop of pipework, or by the simple expedient of putting the lighter components of the blend into the tank last and at the bottom, ie, beneath the heavier ones, so allowing them to float up through the heavier components, mixing on the way.

Organisations such as trading companies that blend away from refineries normally do so in tank farms or on board tankers (sometimes storage vessels, but also seagoing transport vessels). The components that they blend are often purchased from many different companies and countries before being assembled at the blending location. Mixing is achieved by one of the methods discussed in the previous paragraph. On occasion, a blend will be made aboard a tanker which collects the components from different ports or terminals. This requires care, planning and careful implementation. It is not generally acceptable simply to load separately, perhaps into different tanks, components which, if mixed together, would make an on-specification blend. The components must be mixed and the blend homogeneous. For example, the buyer may choose to discharge the cargo at more than one port, so it is not safe to assume that the cargo will be homogenised during discharge.

Blending procedures must therefore be implemented on board from the start. Loading the lighter components last (again, into the bottom of the tanks) may be necessary, as well as leaving some empty cargo stowage space to allow for a pump around. Further full cooperation by the vessel's owners and crew is likely to be needed and this must be agreed in advance. Blending on board must also be performed in accordance with international conventions, such as the International

Convention for the Safety of Life at Sea (1974) adopted by the International Maritime Organization (IMO), which prohibits the blending of bulk liquid cargoes and implementation of production processes on board ships during the sea voyage.[10]

4.3 Blending problems

There are a number of other issues that require consideration, apart from those of blend design and implementation discussed above.

First, blend design is a theoretical process. The predicted results are not always achieved. For this reason, a laboratory-scale physical trial blend should always be made and tested against all required specification parameters, using the actual components to be mixed in the design proportion. Only once this has been checked can the design be regarded with confidence as being suitable.

Secondly, there may be quality issues beyond the available analyses of the components and the required product. For example, when blending heavy fuel oils (marine bunker fuel, 6 Oil etc.) the intended components may not be compatible: that is, they may precipitate heavy, solid asphaltenic material when mixed together, making them unsuitable for use. Compatibility tests, unlikely to be included in the component analytical reports, should be performed before a blending plan is finalised. As another example, components intended for blending into a finished marine fuel may contain chemical wastes for which they have not been tested, though such are forbidden in these grades. The presence of these wastes may not be revealed by the specification analysis. Again, components intended for gasoline blending may contain oxygenates, not identified in their specification analyses but forbidden in the intended final blend. Further, gas oils may have to meet strict colour specifications, whereas intended blend components may be colour unstable. In this circumstance, the colour of a designed blend may significantly worsen with time, despite the design colour having been satisfactorily achieved in the original mix.

Avoiding such difficulties requires a detailed knowledge of the characteristics of the planned blend components which may need to go beyond the initially available analytical data.

5. Summary

Oil trading disputes often arise as a result of an alleged change in the quantity and/or quality of an oil cargo over the course of a shipment. Such disputes can arise due to actual physical loss and/or contamination of an oil. As the preceding sections have discussed, however, such changes may often be caused by factors that could give rise to a claim against a party other than the one that is, at first sight, apparently liable. A careful and detailed investigation will therefore be required in order to establish properly the nature of the losses involved and to allocate responsibility for their cause.

---

10 See Chapter VI of SOLAS Consolidated Edition 2014, regulation 5(2).

Bills of lading

Paul Aston
Holman Fenwick Willan, Singapore

"I, indeed am the Lord of the world, but the law is Lord of the Sea"[1]

1. The evolution of the bill of lading

> *In the Medieval Ages shippers (usually the owners of the goods) as a rule accompanied their cargoes on the voyage to destination and [a] bill of lading served only as an invoice of the goods shipped. [A] [r]ecord of the goods loaded was kept by [the] ship's clerk in the 'Register', 'Book' or 'Writing'. All these names apparently describe a document which was kept by [the] ship's clerk with a copy given to the merchant as a receipt. Such rudimentary bills of lading was [sic] not transferrable, [sic] mainly, because merchant[s] travelled with their goods and while [the] original was in [the] possession of the shipowner the only one copy of bill, held by [the] merchant, was necessary. ... Dealing with masters of ships the '*Extrait du Statut de Sassari de 1316*' says that the masters of ships which intended to leave the port of Torres with the goods of local merchants, shall give a sufficient written security (*de dare suffitiente securitate*) that the merchandise which they have promised to transport shall be shipped in entirety on their vessels. When this written security shall have been given to the merchant in [the] form of [a] '*police de chargement*' (sa puliza) it would allow the ship to leave the port. In French translation '*police de chargement,*' means Bill of Lading, so the record clearly mentioned a document which the shipper received from the master endorsing receipt and warranting shipment of the goods.*
>
> *Later, in the sixteenth and seventeenth centuries, when larger ships ha[d] begun to carry varied cargoes belonging to several shippers this practice gradually came to naught when merchants ceased to travel with their goods and simply dispatched them to a consignee. From this stage [the] bill of lading began to exist as a separate and distinct document to provide a proof that the person demanding delivery of the goods at the port of destination was the person entitled to do so. Naturally, a copy of the register signed by the master or his mate on receipt of goods was such a document to pass a title and bind shipowner and consignee to the conditions of shipment.*[2]

This quotation, based upon commentary penned by legal academic WP Bennett just over a century ago, provides a charming and historical backdrop to the evolution of the bill of lading ('B/L'). The B/L is one of the most ingenious inventions of

---

1 The Emperor Antonius (138–161 AD/ACE).
2 Law and Sea, *Bill of Lading: historical background and short overview* (www.lawandsea.net/COG/COG_Bill_of_Lading.html), drawing on WP Bennett, *The History and Present Position of the Bill of Lading as a Document of Title to Goods* (1914, Cambridge University Press).

commercial life. It may be compared to the invention of paper money, to which it has some similarities. It changed the way business was done and opened up many new commercial possibilities. Like paper money, its popularity has endured.

Initially, B/Ls originated in the fourteenth century as non-negotiable receipts for goods loaded, issued to a merchant who did not wish to travel with the goods. Later, it was found useful to note on the B/L some of the terms of the contract of carriage. In the eighteenth century, the B/L acquired its third function, that of a document of title to the goods.

A very important fourth function has also developed in modern times. The B/L is usually one of the critical documents that triggers payment under a letter of credit. An international 'sale of goods' is in truth a 'sale of documents'. For example, a buyer in China wants to import oil products from the Middle East and enters into a contract of sale with an English seller. The buyer wants to make sure that he does not pay until he is certain, firstly, that the seller will perform his obligations to arrange shipment of the correct goods and, secondly, that the goods are destined to reach his designated port in China. Likewise, the seller has no desire or intention of allowing the goods to leave England without some assurance of payment. This is, of course, where the letter of credit ('L/C') comes in, being established by the buyer's bank in favour of the seller. in order to trigger payment under the L/C, however, the seller must present through banking channels to the buyer's (or confirming) bank documents which conform (ie, those documents required in the credit). Of these, the B/L is a critical document because it evidences both the date of shipment and on which vessel the goods have been placed.

2. Types of bill of lading

There are four types of B/L:

- A negotiable B/L will be consigned "to the order of ..." the shipper or a named party. This may be transferred by delivery and indorsement.
- A bearer B/L will be consigned "to bearer" and may be transferred by simple delivery.
- A straight B/L will be consigned to a named consignee and is not transferable; only the named consignee can acquire rights under it. Under the UK Carriage of Goods by Sea Act 1992 (CoGSA or 'the 1992 Act'), a straight bill falls within the definition of a 'waybill'. The two documents are, however, distinct. For example, a straight bill of lading must be produced in order to obtain delivery of the goods, and is subject to the compulsory application of the Hague-Visby Rules.
- A sea waybill. Sea Waybills are used in lieu of a B/L for straight consignments whenever a letter of credit or similar banking arrangement is not involved in the sale of goods. They are used when cargo will not be traded/sold during transport. They are suitable for regular shipments between related companies which do not require settlements through banks or third parties. They are also used when payment of goods is made under an open account or there is a high degree of trust between the importer and exporter, such that a negotiable transport document is not required under a letter of credit.

Possession of an appropriate B/L will transfer the symbolic possession of the goods. A transferable bill of lading has been described as "the key which, in the hands of the rightful owner, intended to unlock the door of the warehouse, floating or fixed, in which the goods may chance to be".[3] The shipowner has a responsibility to deliver the goods to their true owner. If he delivers to the wrong person, he may be liable to the true owner for the full value of the goods. He will also lose protection and indemnity ('P&I') insurance cover. The shipowner is not, however, privy to the various sale contracts. Presentation of the original B/L allows him to ascertain the true owner. This may be contrasted with the position in respect of a sea waybill, where the true owner only needs to show proof of identity and does not need to present the original sea waybill.[4]

3. The bill of lading as a receipt

3.1 Introduction

A B/L is used in the vast majority of international trade transactions. The B/L:

- evidences shipment of the goods;
- enables the goods to be traded afloat;
- triggers payment for the goods under sale contracts and credit support agreements; and
- enables delivery of the goods to be effected.

Most B/Ls are what are known as 'shipped' B/Ls. This basically means that the B/L will be evidence of the cargo at the time of loading. The date inserted in the B/L will therefore be considered to be the date of shipment and this may have important implications. For example, the value of the cargo in the contract of sale will usually be based on the market value of the cargo on the date shown in the B/L. Where 'received for shipment' B/Ls are issued, these will usually either be returned to the carrier and replaced by 'shipped' B/Ls or be made shipped B/Ls by inserting the name of the carrying ship and the date of shipment.

The consequences of inaccuracy can be severe. It is very important that the B/L is signed and dated accurately to record the actual date on which the cargo was loaded. If the B/L is ante/post-dated, there are serious consequences for the company, which may be exposed to claims from cargo interests. P&I cover may not be available for such claims as the P&I rules exclude cover for claims in certain circumstances where the B/L is ante/post-dated. Charterers and/or shippers may try to persuade the ship's master to accept a letter of indemnity or similar undertaking in return for issuing ante/post-dated B/Ls. Such requests should be resisted due to the risks involved in accepting such letters or undertakings.[5]

---

3 *Sanders Brothers v Maclean & Co* (1883) 11 QBD 327.
4 *APL Co Pte Ltd v Voss Peer* [2002] 4 SLR 481.
5 See 9 below.

3.2 The function of the bill of lading as a receipt

This function was recognised by the common law, and modified by the Hague Rules and the Hague-Visby Rules.[6] The B/L is a receipt as to:

- the quantity loaded;
- the apparent order and condition of the goods; and
- the identifying marks (sometimes called 'leading marks').

The B/L can itself be the contract of carriage, mere evidence of the contract of carriage, or of no contractual significance. As between shipper and carrier, where the shipper is also the charterer, the B/L is not of contractual significance as the charterparty is the contract of carriage. Where the shipper is not the charterer, the B/L is evidence of the contract of carriage, but is not itself the contract because the contract would have been entered into before the B/L is issued. Where the contract of carriage has been concluded before the B/L is issued, the carrier cannot unilaterally alter the terms of the contract by introducing contradictory terms in the B/L. When and how the contract of carriage is entered into depends on the circumstances of each case.

The first person to receive the B/L is the shipper, he who loaded the cargo. The shipper, because he loaded the goods, should have some knowledge both of the quantity loaded and their condition. Therefore, in a dispute between the shipper and the carrier, the statements in the B/L are only *prima facie* evidence of the quality and condition of the goods loaded. This means the carrier is free to bring evidence to contradict the B/L and the shipper can retaliate with evidence to uphold it.

Where the B/L is negotiated to a subsequent holder, however, it becomes conclusive evidence of the quality and condition stated. As between a third-party B/L holder and the carrier, the B/L constitutes the contract of carriage between them. The terms of the B/L are the only contractual terms which govern the relationship between the carrier and a third-party holder of the B/L. At common law, this happens by means of an estoppel: the holder has to prove detrimental reliance on the statement in the B/L. Under the Hague-Visby Rules, the statements are automatically considered conclusive evidence against the carrier, once the bill is transferred to a third party in good faith.[7]

It should be noted that the carrier does not certify the quality of the goods, but only their "apparent order and condition". The carrier is only bound to note on the bill defects that would be apparent on superficial inspection of the goods by a reasonable ship's master, relying on his general knowledge of carriage of goods. Thus, if bags of rice are loaded wet, stained or torn, that should be noted on the bill. If, however, the rice inside the bag is mouldy or discoloured, or simply not the type of rice described in the bill of lading, that is not something which the carrier would be expected to discover and note on the bill. The master is not meant to adopt or usurp the role of the quality inspector.

---

6 More is said about the rules at 9 and 10.1 below.
7 Article III.4.

4. The bill of lading as a contract of carriage

Discussion now turns briefly to the function of the B/L as evidence of the contract of carriage. It should be noted first that the B/L is not called "the contract of carriage", but only "evidence of the contract of carriage". This is a matter of timing. The bill is only issued when loading is complete. As mentioned above, however, the arrangement to load – in law, the contract – must inevitably come before that, as one does not load one's goods on a ship and only then open negotiations for a contract of carriage. Usually there is little evidence of the negotiations and terms agreed before shipment; the B/L, although issued later, is then the best evidence of what was agreed.

Charterparty bills of lading can be compared to train tickets containing a statement on the back saying "for full terms and conditions please ask for a leaflet at the ticket office". The full terms are not included in the bill, but the holder is referred to another document in which they are set out – the charterparty. Often, but not always, the B/L contains some of the terms: typically, a clause paramount incorporating the Hague-Visby Rules. For the full terms, however, it is necessary to refer to the charterparty.

5. Incorporation by reference

Bills of lading rarely expressly include a choice of law clause. Most commonly, and assuming a sufficiently wide incorporation clause, the law and jurisdiction provisions of the charterparty will therefore apply. Arbitration clauses are very common for disputes arising in connection with charterparties and B/Ls.

A few points must be noted about charterparty bills:

- Only parts of the charterparty relevant to loading, carriage and discharge of the goods will be incorporated. Thus, if a time charter is incorporated, the clauses relating to payment of hire and bunkers, for example, will not be incorporated.
- A charterparty arbitration clause will not be incorporated unless the B/L specifically refers to it.
- If the shipper is also the charterer, the B/L has no contractual effect as between the carrier and the shipper. Their contract remains the charterparty. The B/L will, however, acquire contractual effect between the carrier and the subsequent holder of the bill.
- Full terms are not stated in the B/L. Instead, the B/L incorporates the terms of a charterparty (which is usually identified by date). Therefore, it is generally good practice to ask for a copy of the charterparty. Problems can arise where there is doubt over which charterparty is incorporated and over which charterparty terms are incorporated. This is a question of construction of the incorporation clause, but some presumptions apply. Which charterparty is incorporated? Issues arise where there are several charterparties in a chain and where either no date is inserted on the face of B/L, or where multiple charterparties have the same date. The first presumption is that it is the head charterparty which is incorporated.[8] This is

8 *The San Nicholas* [1976] 1 Lloyd's Rep 8.

> because the shipowner is a party to that charterparty. Where, however, the charterparty chain includes a voyage charterparty underneath a time charterparty, the voyage charterparty is incorporated.[9] This is because the voyage charterparty covers the specific voyage to which the B/L relates.

This area still gives rise to disputes that may find their way to court. In a recent case,[10] the B/L stated on its face, "Freight payable as per CHARTER-PARTY dated 11/04/13" and included the following term on its reverse:

All terms and conditions, liberties and exceptions of the Charter Party, dated as overleaf, including the Law and Arbitration Clause, are herewith incorporated.

The charterparty referred to on the face of the B/L was a time charterparty and included, so far as material, the following clauses:

17. Bimco Arbitration Clause to apply – as per Clause 93 – with English Law and arbitration in London...Any dispute arising out of this Charter Party and from the issuing of Bill of Lading shall be referred to arbitration in London and English law to apply.

A further provision stated that all B/Ls issued under the charterparty were to incorporate the arbitration clause and English law. Arbitration was commenced under both the charterparty and the arbitration clause. Despite this, the Chinese cargo interests instituted proceedings in People's Republic of China. The claimant sought to restrain the defendant from litigating matters under the B/L otherwise than in the London arbitration. The claimant also sought an interim injunction, which was granted by Blair J on April 24 2015. Blair J granted permission to serve the defendant out of the jurisdiction and by way of alternative service. The defendant opposed the continuation of the injunction on three grounds. The first and the third points are the only ones that are relevant for present purposes. The first point raised arguments concerning the incorporation of the charterparty and the arbitration clause contained therein. The third point contended that there were strong reasons not to grant the order as a matter of discretion, even if the arbitration clause were to be treated as incorporated and the court did otherwise have jurisdiction.

The claimant accepted that the B/L referred to the incorporation of the charterparty dated April 11 2013 and that it incorporated all terms and conditions of the charterparty "dated as overleaf", including the law and arbitration clause. It submitted, however, that it was arguable that, on a true construction, the reference to a charterparty dated April 11 2013 should not be read as referring to the time charterparty of that date but to a voyage charterparty of November 28 2013, which was not referred to expressly anywhere in the B/L. In support of that contention, the claimant referred to the fact that the B/L referred to freight being payable as per the specified charterparty, whereas a time charterparty such as the April 11 2013 charterparty provided for the payment of hire rather than freight. The claimant relied upon an authority,[11] in which the Court of Appeal had held that, where

9 *The Nanfri* [1979] AC 757.

10 *Southport Success SA v Tsingshan Holding Group Co Ltd* [2015] EWHC 1974 Comm.

11 *The SLS Everest* [1981] 2 Lloyd's Rep 389.

reference to a charterparty in a B/L was left blank, it was readily to be inferred that a voyage charterparty was intended to be incorporated. The court in that case said, however, that it was clear that there was no bar to the court finding that a time charterparty was incorporated by reference in, for example, a B/L. In a more recent case,[12] Mr Jonathan Hirst QC, sitting as a judge of the High Court, stated as follows:

> *The bills of lading expressly provided that all terms and conditions, liberties and exceptions of the Charter Party dated September 28 2005 'including the Law and Arbitration Clause' were incorporated. The charterparty was a time charter rather than a voyage charter, so hire rather than freight would be payable. There can be no doubt that the parties intended the terms of the time charterparty (to which the sellers Capezzana were party) to be incorporated in the bill of lading contracts. The express reference to the arbitration clause would be sufficient to incorporate a charter party arbitration clause, even if it required a degree of manipulation ...*

In the court's judgment in the present case, the reference to the charterparty of April 11 2013 was entirely clear and unambiguous. There was nothing impermissible or necessarily irreconcilable about the incorporation of the terms of a time charter and the court saw no reason to doubt that the time charterparty of April 11 2013 was incorporated as per the express terms of the B/L.

This case is salutary in that it shows the potential for disputes on matters of incorporation by reference.

6. The parties

Who is party to the B/L contract? From the cargo interests' perspective, this tells that party who can sue the carrier under the B/L (ie, a claim for breach of the contract of carriage). If it is not possible to sue under the B/L, the claim can possibly be brought in tort (ie, for negligence) instead of under contract by the owner of the cargo.

The question of which party can sue the carrier under the B/L depends on the type of carriage document. With transferable/negotiable ('to order') B/Ls, the shipper's rights of suit are transferred to the lawful holder (ie, the consignee or the endorsee) once the B/L is transferred. Under sea waybills and delivery orders, the shipper can sue (as the original party) as well as the party/parties named in a sea waybill or delivery order.

With transferable B/Ls, the scenario may arise where the party who suffered the loss is not the lawful holder/the person entitled to sue. The right to sue belongs to the lawful holder (eg, the consignee); the lawful holder has, however, transferred title/property over the goods to a third party (its customer) and it is the third party who has suffered the actual loss. In such a case, the lawful holder holds the right to sue for the benefit of the person who has actually suffered the loss.

7. Transfer of rights under a bill of lading

What is a document of title? It is a document issued by a named person, giving the holder or bearer of that document the right to demand certain goods or money from that person. The characteristic of negotiability allows the document, and the rights

12 *The Duden* [2009] 1 Lloyd's Rep 145.

that go with it, to be transferred from person to person in the course of commercial transactions.

Typically, a B/L is issued to the shipper, and then transferred by the shipper to one or more further parties, eventually being transferred to the receiver. The original contract of carriage is made with the shipper. At common law, the subsequent holders of the B/L did not acquire any contractual rights against the carrier. This was inconvenient because, generally, at the time the goods reached their destination, the B/L would have been transferred to another party and the goods would have been at that party's risk. The Bills of Lading Act 1855 was enacted to address this situation. The basic scheme of that Act was that when the B/L was transferred to a party other than the shipper, it was also effective to transfer rights of suit to that party. The 1855 Act has since been replaced by the CoGSA 1992, which follows the same scheme. In particular, section 2(1)(a) of the 1992 Act provides that:

> *a person who becomes – (a) the lawful holder of a bill of lading ... shall (by virtue of becoming the holder of the bill) have transferred to and vested in him all rights of suit under the contract of carriage as if he had been a party to that contract.*

Thus, as the B/L is transferred from party to party, so also is the right to sue the carrier. It is also important to note that where rights are transferred under the 1992 Act to a lawful holder of the B/L by the shipper, the shipper's rights under the contract of carriage are extinguished. Where the B/L is subsequently transferred from one lawful holder to another, the rights of the previous lawful holder are extinguished. The effect of this is that only the last lawful holder in the chain has title to sue. For this reason, a claim brought in the name of a shipper is likely to fail if the shipper has transferred the B/L to another party.

Possession of an appropriate B/L will transfer the symbolic possession of the goods. A transferable bill of lading has been described as "the key which, in the hands of the rightful owner, intended to unlock the door of the warehouse, floating or fixed, in which the goods may chance to be".[13] The shipowner has a responsibility to deliver the goods to their true owner. If he delivers to the wrong person, he may be liable to the true owner for the full value of the goods. He will also lose P&I insurance cover. The shipowner is not, however, privy to the various sale contracts, so thtat presentation of the original B/L allows him to ascertain the true owner. This must be contrasted with the position in respect of a sea waybill, where the true owner only needs to show proof of identity and does not need to present the original sea waybill.[14]

8. The identity of the contracting carrier

A further important point to make is that it is not always the shipowner who is the contracting carrier under the B/L. There may be a 'non-vessel-owning carrier' or NVOC. As the name suggests, an NVOC is a person who enters into a contract of carriage as carrier, but without being the owner of the vessel. It is therefore important to consider the B/L carefully in order to identify the contracting carrier. Particularly important in this respect is the signature box on the front page. One clear and

13 *Sanders Brothers v Maclean & Co* (1883) 11 QBD 327, per Bowen LJ at 341.
14 *APL Co Pte Ltd v Voss Peer* [2002] 4 SLR 481.

reliable rule here is that the signature of the master is effective to bind his employer, the shipowner; and equally, the signature of an agent 'on behalf of' the master. If a particular party is named as the 'carrier' in the form of signature, or the bill is signed by or on behalf of a particular party, then it is likely that that party will be the carrier. Some consideration should also be given to 'identity of carrier' clauses on the reverse of the bill. Further, the document must be considered as a whole, and given reasonable business sense. If the identity of the carrier is manifest on the face of the bill, then it is unlikely that this identification will be contradicted by a small print clause on the reverse.[15]

9. The charterer's rights and indemnities

The Hague-Visby Rules require that, after receiving the cargo, the carrier or his master or agent must, on the shipper's demand, issue a B/L to that shipper. More detailed provisions relating to the shipowner's obligations with regard to issuing a B/L will be set out in the charterparty. For example, BPTIME 3 provides that "Bills of lading and waybills shall be signed as Charterers direct, without prejudice to this Charter". This obligation to sign B/Ls 'as directed' relates to the form of the B/L, not the facts it states: it does not require or condone the signing of B/Ls that are inaccurate. A carrier cannot, for example, be obliged to sign a B/L that misstates the quantity or quality of the cargo. The charterer's right to direct the form and content of the B/L is balanced by the indemnity included in the charterparty by which the charterer agrees to indemnify the shipowner "against all liabilities that may arise from signing the B/L and waybills in accordance with the directions of Charterers to the extent that such B/Ls and waybills impose more onerous liabilities than those assumed by Owners under the terms of this Charter".[16]

Both the Hague Rules and the Hague-Visby Rules provide that the description of the cargo in a B/L (weight, condition, etc) "shall be *prima facie* evidence" of the correctness of the description so given. Under the Hague-Visby Rules, where the B/L is transferred to a third party in good faith, the statements describing the cargo become conclusive because "proof to the contrary shall not be admissible".[17] It can, however, be difficult for the carrier to ascertain the condition of the cargo and there can be genuine debate about the exact amount of cargo on board. It is for this reason that most B/Ls include a phrase such as "weight and condition unknown" and "apparent good order and condition" (or an equivalent). By the inclusion of these provisions, a holder of the B/L will not be entitled to claim against the carrier for any shortfall or shortcomings by basing his claim on the B/L particulars. These provisions are, however, only effective in cases of genuine doubt, or where checking is not possible. They cannot be used to cover known discrepancies or discrepancies that could readily have been detected.

This can give rise to huge problems in the oil trade, and disputes between shippers/charterers and owners' representatives (masters, officers, surveyors,

15 *The Starsin* [2003] 1 Lloyd's Rep 574.
16 This example is taken from BPTIME 3.
17 Hague-Visby Rules, article III, rule 4.

superintendents) over whether the cargo quality to be shipped justifies a clean B/L or whether the master must clause the B/L, otherwise he will be in danger of misrepresenting the "apparent good order and condition" of the cargo. Given that clean B/Ls are invariably required under letters of credit, one can readily appreciate the pressure masters and owners are placed under to issue them. Their problem is compounded by the fact that any letter of undertaking by the shipper or the charterer to protect the master/owners against the consequences of issuing a clean B/L is, as a matter of English Law, almost certainly going to be held unenforceable.[18] In addition, the master may well prejudice his right of indemnity under the charter in signing B/Ls as presented. The failure to clause the B/L properly will most likely be treated as a *novus actus interveniens* in the chain of causation.[19] Courts take the view that such letters of indemnity facilitate a fraud, ie, a deliberate act of inserting into the B/L information known to be incorrect. Such letters of indemnity are not legally binding, and therefore offer the Company no protection if the shipper goes back on his promise. No P&I cover will be available either, because P&I Clubs exclude cover for claims arising from the issue of an ante/post-dated B/L or claims arising from the issue of a B/L known by the master or the club member to contain an incorrect description of the cargo, its quantity or condition.

This is a very complex area but fortunately does not require unpacking because, in the tanker trade, the master is never required to clause a B/L based on the "apparent good order and condition" of the cargo. The nature of the carriage of hydrocarbons is that they pass through pipes and manifolds, from shore tanks into ship's tanks. Their quality on loading is the subject of inspection, sample taking and analysis, all of which are dealt with in detail elsewhere.[20]

10. International conventions

A series of conventions provide a legal regime governing the rights and obligations of shippers, carriers and consignees. As noted in a well-known guide:[21]

> *The intention of the conventions is to provide a balanced regime of rights and obligations so that the Cargo Owner can expect a certain standard of care from the Shipowner in maintaining his ship and in caring for the cargo; and the Shipowner can be protected from claims where, despite his maintenance and care, the cargo is damaged.*

Their terms are compulsorily applicable to B/Ls issued in a country that has implemented one of these conventions in its own domestic law; any provisions of a B/L that are inconsistent with the applicable onvention are deemed void. The three key conventions which are currently in effect are:

- the Hague Rules 1924 (to which the US is a key contracting state);
- the Hague-Visby Rules 1968 (which are in force in the majority of key trading nations); and
- the Hamburg Rules 1992 (which have had very limited uptake).

18 *Brown Jenkinson & Co Ltd v Percy Dalton (London) Ltd* [1957] 2 QB 621.

19 *The Nogar Marin* [1988] 1 Lloyd's Rep 412.

20 Editorial note: See chapters 'Quality and quantity determination' and 'Practical aspects of quantity and quality determination in relation to shortage and contamination claims'.

21 Stephen Mills, *Bills of Lading: A Guide to Good Practice* (1998, North of England P&I Association).

10.1 The Hague Rules and the Hague-Visby Rules

The Hague Rules were adopted by a large number of countries, including England & Wales (through the enactment of the Carriage of Goods by Sea Act 1924) and the USA (through the enactment of the Carriage of Goods by Sea Act 1936, which is still frequently incorporated contractually into B/Ls and charterparties). By the 1960s, however, there was felt to be a need to update the rules and to amend certain provisions that were causing particular problems. Following a meeting at Visby in Sweden, the Hague Rules were amended by the Brussels Protocol 1968 and many countries updated their internal legislation to give effect to the Hague-Visby Rules.

The Hague-Visby Rules have been developed for the protection of the holder of the B/L, who is not generally a party to the original contract of carriage. The rules impose on the carrier obligations in relation to the standard of the ship, cargo areas and crew, as well as on the manner in which cargo operations and care should be conducted.

The basic formula for application of the Hague-Visby Rules focuses on the document covering the contract of carriage rather than the contract of carriage itself. Article X of the Hague-Visby Rules provides that they apply to every B/L relating to the carriage of goods between ports in two different states if:

- the B/L is issued in a contracting state; or
- the carriage is from a port in a contracting state; or
- the contract contained in or evidenced by the B/L provides that the Hague-Visby Rules or the legislation of any state giving effect to them are to govern the contract.

Under article III.1, carriers are obliged to "exercise due diligence before and at the beginning of the voyage to make the ship seaworthy". This 'seaworthiness' obligation is non-delegable, albeit that 'due diligence' replaces the absolute and ongoing seaworthiness obligation under the common law. The term 'seaworthiness' covers:

- the physical condition of the ship;
- the suitability of the storage area and facilities for the cargo;
- the manning and management of the ship; and
- documentation and certification of the ship.

The carrier is also obliged by article III.2 to "properly and carefully load, handle, stow, carry, keep, care for and discharge the goods carried".

Claims against the carrier can be defeated by the "excepted perils" set out in article IV.2, which lists a number of perils for which neither the carrier nor the ship shall be responsible for loss or damage. The most common defences are:

(a) Act, neglect, or default of the master, mariner, pilot, or the servants of the carrier in the navigation or in the management of the ship.

(b) Fire, unless caused by the actual fault or privity of the carrier.

...

(m) Wastage in bulk of weight or any other loss or damage arising from inherent defect, quality or vice of the goods

In the *Iron Gippsland* case,[22] damage occurred to a cargo of Singapore Gas Oil, also known as 'ADO'. ADO is an inflammable product; therefore, to prevent ignition of the explosive gas, the tanks are first filled with inert gas. The ADO was contaminated via the inert gas system and the question to be answered in the proceedings was whether the contamination of the ADO either was the result of failure to care for the cargo in accordance with article III.2 of the Hague-Visby Rules or fell within the scope of article IV.2(a) and was therefore an act covered by the nautical fault exception. In the Supreme Court of New South Wales, Australia, Carruthers J construed the exception strictly and said that:

> *It is true that inert gas systems were installed on tankers fundamentally for the protection of the vessel. However, the purpose of the inert gas system is primarily to manage the cargo, not only for the protection of the cargo but for the ultimate protection of the vessel from adverse consequences associated with that cargo and, in my view, damage occasioned to cargo by mismanagement of the inert gas system cannot be categorized as neglect or fault in the management of the ship.*

Even though it was accepted that the system existed fundamentally for the protection of the vessel, it was held that this was achieved through the management of cargo and accordingly this was not an act in the management of the vessel.

The case of *The Fiona*[23] involved an explosion in a cargo tank when a surveyor inserted into it an unearthed temperature probe. Explosive gases had built up inside the cargo tank, resulting partly from a cargo of fuel oil and partly from residues of the previous cargo which the owners had failed to wash out properly before loading. The owners failed in their claim for an indemnity from the shippers of the fuel oil in respect of loss and damage caused by the explosion, notwithstanding the fact that the fuel oil itself was deemed to be dangerous due to the fact that it emitted flammable gas. It was held that the shipowner's failure to wash away the residues of the previous cargo had contributed to the explosion and therefore they were unable to discharge their overriding obligation of seaworthiness under article III.1, precluding them from recovery under article IV.6. The case established that where the breach of the seaworthiness obligation is an effective cause of the damage, then the shipper will be liable for all damage at any time without any liability to the owner. The shipper is always liable to the shipowner where damage is caused by dangerous goods, regardless of whether the shipper still has cargo interest.

10.2 The Rotterdam Rules

The Rotterdam Rules[24] were recently adopted but have yet to come into force. They provide for 'electronic transport records', clarify that these documents may be negotiable or non-negotiable and prescribe that they should be treated largely in the same way as paper documents. The Rotterdam Rules will have a far broader

---

22 *The Iron Gippsland* [1994] 1 Lloyd's Rep 335, 358.
23 *The Fiona* [1994] 2 Lloyd's Rep 506.
24 Made pursuant to the United Nations Convention on Contracts for the International Carriage of Goods Wholly or Partly by Sea 2009. When in force, this convention will replace the UN Convention on the Carriage of Goods by Sea 1978, which implemented the Hamburg Rules.

application than the Hague-Visby Rules and the carrier's obligation will be to "make and keep the ship seaworthy" instead of just to "make the ship seaworthy".

11. Tanker bills and problems specific to the tanker trade

Tanker B/Ls tend to have few terms on their reverse side. These usually consist of:

- a 'Clause Paramount' (which contractually incorporates the Hague Rules or Hague Visby Rules if they do not automatically apply);
- a 'General Average' clause (which specifies the general average regime that will apply in case of dispute);
- a 'New Jason' clause (regarding the parties' obligations to contribute to general average); and
- a 'Both-to-Blame Collision' clause (which deals with cross-claims and counterclaims in the event of a collision).

Other clauses, depending on the trade route, may include 'War' clauses, 'Ice' clauses and 'Strike' clauses.

11.1 Consequences of misdelivery or delivery without production of an original bill of lading

The consequences of misdelivering or delivering without production of an original B/L are serious. The carrier is likely to be held fully liable for wrongful delivery of the cargo, with the consequence it will have to compensate the rightful cargo owner for the full value of the cargo – this could amount to a substantial sum. P&I cover may not be available as P&I Rules exclude cover for claims in certain circumstances where proper delivery has not been made.

That said, the fact remains that almost every cargo of oil is delivered without the production of any B/L and that this is the norm rather than the exception. This is reflected in the fact that almost every tanker charterparty will require the owner to agree to discharge without the production of a B/L in return for a Letter of Indemnity ('LoI') issued by the charterer. This LoI will normally be in a wording approved by the major P&I clubs. Even though doing so will prejudice P&I cover, the P&I clubs understand that this is the industry practice and accordingly suggest LoI wordings that will provide appropriate protection for the carrier. It is important to note that an LoI against the consequences of delivering cargo without a B/L is not tainted in the same way as an LoI against the consequences of issuing a clean B/L. There is no question of any misrepresentation in the former case, whereas there may well be in the latter.

The indemnity is, however, only ever as good as the party who provides it and such LoIs will often not be counter-secured by a bank, for obvious reasons of commerciality and cost. If the issuer is a major oil company, there is unlikely to be a problem, given the balance sheet of the company giving the LoI. This may not, however, be the case with an independent trader, who may be known to go into insolvency or simply refuse to honour the LoI by raising defences that it did not have authority or capacity to offer any such indemnity under its articles of association or memorandum of association or pursuant to the procedural law of its place of incorporation. An example of such an objection would be that to honour such an LoI may place it in breach of the law of its place of residence or incorporation, eg, where the relevant regulatory approval (eg, State Administration of Foreign Exchange (SAFE) in the PRC) was not obtained. The lesson to be drawn from this is that there is no point too bad for some corporations to take where they are faced with a demand to pay out under their LoI in a bad market.

Frequently, where there is more than one charter party but only one vessel and one cargo, there will be a number of LoIs passing up and down the line. It will therefore be necessary to ensure that an LoI is issued by someone whose bank balance can handle it if and when a demand is made because the party who owned the cargo or held the B/L did not receive the cargo.

A well-known industry publication notably said the following on this topic in the 'Clause Notes' in the 1989 edition when discussing the B/L clause in tanker charters:[25]

> *[I]n recent times the most burning issue in terms of the content of the clause has been whether charterers should be entitled to require discharge without production of an original Bills of Lading... The majority of regular tanker charterers now expect a clause requiring owners to discharge without production of an original Bill of Lading...It requires to be borne in mind that delivery without production of a Bill of Lading is viewed in many legal systems as a default of the utmost gravity. In such circumstances the governing law of the charterparty will be of no avail, as it is the attitude of the local court where the ship is detained that has to be reckoned with.*

11.2 Early departure procedure/signing blank bills of lading

Predominantly in the tanker trade, a shipper, loading terminal or charterer may request the ship to follow Early Departure Procedure ('EDP'). Among other things, EDP usually involves the master issuing a signed but otherwise blank B/L form. Alternatively, the B/L may be completed except for the quantity or weight. Clearly, this procedure exposes the Company to significant liabilities (see 3.1 above). Accordingly, if the master is requested to follow EDP and/or sign a blank B/L, he should refuse and contact his managers immediately.

11.3 Delivery of cargo against a bill of lading retained on board

It has been known for a shipper to pressurise the master to retain on board one original B/L out of the set (usually three) issued, with the instruction to hand it to the intended receiver in order to avoid the possibility of there being no original B/L for presentation at the discharge port. The intended receiver presents the original B/L back to the master and claims delivery of the goods. The risk involved in this practice is that since the master knows or ought to know that, through ordinary commercial channels, competing claims for the cargo can arise under the original B/L not retained on board, he will be deemed to have acted in bad faith by delivering the cargo against the original B/L retained on board. If a competing claim does arise, the

25 Harvey Williams & Stan Bonnick, Commentaries on *Tanker Voyage Charter Parties* (1989, International Association of Independent Tanker Owners).

carrier may be held fully liable for wrongful delivery of the cargo and for compensating the rightful cargo owner for the full value of the cargo. As stated previously, P&I cover may be lost for claims in certain circumstances where proper delivery has not been made.

11.4 Commingling or blending cargo on board

Commingling or blending is mostly associated with oil cargoes in bulk. Cargo interests may wish to commingle or blend cargoes loaded on different dates and/or at different places and/or with different specifications. Should this be requested, the master should refuse to perform any commingling or blending until he has received instructions from his managers. The commingling or blending may affect the specification of the cargo already loaded and the carrier may be held liable under any B/L already issued for such cargo. The same liability may arise under the B/L to be issued for the second parcel of cargo to be loaded. While a remark can be inserted in the B/L for this parcel drawing attention to the commingling/blending, should the cargo be loaded in apparent good order and condition, there will be no basis for clausing which casts doubt on the apparent order and condition. There may be complications with regard to dates and places of shipment stated in the B/Ls because, if these are different, a single B/L should not be issued.

11.5 Split bills of lading and delivery orders

Split B/Ls are bills issued for part of a cargo originally shipped under a single set of B/Ls. They are most common in the bulk trades, but are also not uncommon in the oil trades, eg, where cargo interests may wish to split a cargo covered by a single B/L under which delivery can only be made to one party, between, say, three receivers. If such a request is made, all the original B/Ls of the set first issued for the consignment to be split should be surrendered and replacement B/Ls issued. Care must be taken to ensure that the total amounts of cargo stated in the split B/Ls equal the amount in the original B/L, and that the information, eg, the description of the cargo and date, is the same. It is also advisable to clause the replacement bills as there is an increased risk of one or more of the receivers receiving more or less cargo than they are due. Once the replacement B/Ls have been issued, delivery should only be made to the receiver named in the those bills of lading and for the amount stated in the particular B/L. Care will obviously have to be taken to deliver an amount as close as is possible to that stated in the B/L.

Delivery orders may be requested instead of split B/Ls because cargo interests do not require B/Ls that allow them to trade the cargo during the carriage. The person entitled to delivery simply requires the cargo to be delivered at discharge to other parties. The same precautions regarding the use of split B/Ls apply to delivery orders.

11.6 Delivery to a destination not named in the bill of lading

Cargo interests may request delivery of their cargo at a place other than that named in the B/L. It may be that all original B/Ls can be surrendered and replacement B/Ls issued. If, however, this cannot be done, the owners will have to make other arrangements to protect themselves by complying with such a request. There is a risk of an original B/L being presented for the cargo at the place stated in it and, in addition, the carrier could be deemed to have committed a deviation by discharging at a different port other than that stated in the B/L. A deviation is a breach of the contract of carriage and has serious consequences, eg, loss of P&I cover.

12. Trading and chartering

This is an area of considerable topical interest with dot.com start-ups attempting to cut out brokers and allow B2B trading. The recent history of e-trading is, however, more negative than positive, with the market reporting more failures than successes – eg, EnronOnline and Emetra (metals trading). These setbacks are probably illustrative of both traders' and owners' concerns in moving to paperless trade.

12.1 What are the advantages of an e-trade?

- At the actual conclusion of a trade stage, the advantage is speed. A trader can offer a product for sale at a certain price, and another trader can decide to buy that product at the advertised price or seek to negotiate. There are various different mediums through which to do this: either through the use of an internet marketplace or (at its simplest) through the exchange of emails.
- A classic issue in oil trading is that of the daisy chain contracts. These chains can often involve 10 or more parties. The result of this is that when the vessel arrives at the discharge port, the B/L is still making its way down the contractual chain. The result is that the receiver/charterer provides the owner with an LoI agreeing to indemnify the owner in a number of different scenarios, in particular if the owner is later sued for releasing the cargo to the wrong legal entity. So commonplace is this that most charterparties now include a rider clause setting out an agreed wording for an LoI. These wordings are often very standard, having been approved by the International Group of P&I Clubs. The advantage of an e-trade is that this delay can in theory be avoided.
- The cost of couriering the B/L across the world is high. Coupled with the back office staff who are required to deal with the incoming courier, make the necessary checks, liaise with the banks and then send the documents out, there would be a considerable costs saving if there were a central registry where the B/L could be lodged, and each interested party could then pass their legal interest in the cargo by endorsing it to their buyer, without the B/L having to leave the central registry.
- E-trade can reduce the incidence of fraud. The risks of sending a B/L back and forth across the world are well known. By using an e-bill, subject to stringent security measures, this risk can be reduced. It is possible for an agreement to be concluded by the exchange of email messages, or by or through the medium of a market-based website.

Is it therefore possible to replicate a paper trade in the electronic medium? At the discharge port, the lawful holder of the physical B/L must deliver the goods to that holder. Transfer (or assignment) of that bill enables the party to pay for the goods, secure in the knowledge that transfer of title in the goods has been passed to it. Presentation by the shipper (or other party down the line) of the B/L under the letter of credit enables:

- the shipper to be guaranteed payment;
- the consignee (or assignee) to take title in the goods and to take over the shipper's rights in relation to the proper performance of the contract of carriage by the carrier; and
- the bank to control its security over the goods through the medium of possession of the B/L, ie, by the exercise of a lien.

As it stands, English law contains no provisions on electronic transferable records and, as the English courts have not yet recognised and endorsed any mercantile custom in this regard, creating an electronic alternative to the B/L that performs the bill's document of title functions remains problematic under English law. This is the case even though, as discussed at 12.2–12.4 below, practical difficulties in realising an electronic equivalent to a concept so intimately linked to paper have been largely resolved by technological advances. Because the law does not contain the necessary provisions, electronic B/L systems designed to operate under English law must be based on multipartite agreements that effect the desired transfers of rights through the concepts of novation and attornment.

In order to facilitate the development of electronic alternatives to documents of title, it is essential to translate, in legal terms, the concepts of 'possession' and 'holdership' into concepts that make sense in the electronic medium. This translation needs to be understood uniformly in different jurisdictions, as the appeal of documents of title, in particular B/Ls, for cross-border trade is exactly the fact that they are effectively universally recognised as having more or less the same 'powers', regardless of the jurisdiction concerned.

12.2 The SEADOCS project 1986

SEADOCS was the first significant attempt to apply electronic technology to the carriage of goods by sea. The project, managed by London-based SeaDocs Registry Ltd on the joint initiative of Chase Manhattan Bank and the International Association of Independent Tanker Owners (INTERTANKO) had among its primary goals the protection of the victims involved by fraudulent alteration of paper B/Ls in the oil trading business.[26] SeaDocs Ltd attempted to simulate a negotiable B/L using a central registry, held by Chase Manhattan Bank, to record the phases in the passage of property in the shipment.

The shipper would release a paper B/L which SeaDocs Ltd held as an agent for all the parties. An electronic key code would be released to the shipper, who was required to communicate to SeaDocs Ltd electronically its intention to negotiate the bill. The shipper would then provide the buyer/endorsee with a portion of the code; upon SeaDocs Ltd ensuring the accuracy of the information received, the buyer/endorsee would be enlisted into the registry as the 'legal owner' of the cargo.

---

26 George Chandler, *The Electronic Transfer of Bills of Lading* (1980) 20 J Mar L & Com 571.

The buyer/endorsee would therefore be issued with an electronically generated B/L enabling it to take delivery at the docks.

SEADOCS failed for a series of reasons, both practical and legal. Firstly, it was not free from paper documents; rather, it was heavily dependent on them, both on shipping and on delivery, and the $500 cost of each operation was in addition to the costs of the traditional paper system. Secondly, it was difficult to pinpoint risk and liability in the case of an error to any of the parties, with serious repercussions on the cost of insurance. Further, other banks were understandably dissatisfied with one of their main competitors' exclusive access to the registry.[27] The major legal concern was, however, that SEADOCS failed to achieve true negotiability, as the change in ownership involved communication both to the carrier and to the endorsee.

The project demonstrated, however, that the idea of dematerialising negotiable B/L was achievable through a system based on a central registry.

12.3 The CMI model

The Comité Maritime International (CMI) Rules for Electronic Bills of Lading 1990 ('CMI Rules') are an addition to the United Nations Rules for Electronic Data Interchange (UN/EDIFACT). They have no force of law but, unlike SEADOCS, can freely be incorporated in a contract of carriage. Their incorporation would prevent parties from claiming a defence that the B/L is not in writing. While the rules are intended to govern all aspects of electronic B/Ls, rule 6 clearly states that:

> *The Contract of Carriage shall be subject to any international convention or national law which would have been compulsorily applicable if a paper bill of lading had been issued.*

Although the use of the CMI Rules has arguably had an impact as to the manner of handling EDI B/Ls on an international scale, it is evident that the Hague-Visby Rules, for example, will continue to govern negotiable B/Ls. The CMI Rules leave the parties freedom of choice as to the format of messaging to adopt, although rule 3(b) encourages the use of UN/EDIFACT.

The CMI Rules do not provide for the use of a central registry, such as was the case in SEADOCS and as will be seen with regard to Bolero (see 12.4 below). In fact, the working of the CMI model reflects the steps in the paper-based transaction, in which the carrier notifies receipt of the goods electronically to the shipper. However, the system of transfer of the virtual B/L created an area of uncertainty in that carriers would have been privy to each transfer, they had knowledge of the identity of each transferee and carried potentially heavy burdens of liability in the case of a loss due to a failure of the system. The rules failed to make provision for a failure system and lacked reference to what constituted receipt of offer or acceptance.

Although the CMI model was not successful, it played an important role in giving birth to subsequent attempts at electronic B/Ls, such as Bolero.

12.4 Bolero

Bolero[28] has been formed to allow the trade to continue in the e-medium after the

27 ET Laryea, *Bolero Electronic Trade System – An Australian Perspective* (2001) 16(1) JIBL 4–11.
28 www.bolero.net. The name 'Bolero' is an acronym for Bill of Lading Electronic Registry Organisation.

trade has been concluded. Bolero has based itself so far as possible on replicating the paper world, in particular the role performed by the B/L. The two leading lights behind Bolero are SWIFT, the international banking cooperative, and the TT Club, a P&I Club; respectively, they provide expertise in, firstly, secured transmission of banking messages and transfers and, secondly, the appreciation of the role performed by the carrier. Operators wishing to use the system must first join Bolero Associated Ltd (BAL), which is appointed as an agent for the purpose of entering into a contract regulated by the Bolero Rule Book (the 'Rule Book'). The system is governed by English law and gives the English courts jurisdiction in the case of disputes.

The prominent feature of the project is the Rule Book, which gives contractual force to the formal agreement between users and between BAL and each single user. The Rule Book provides a legal framework in the form of a multilateral agreement between all users, who undertake not to challenge the validity of any transaction or communication made by a Bolero signed message on the grounds that it was made in electronic form instead of by paper and/or not signed or sealed. The signatories agree to treat the digital messages as if they were in writing, particularly with regard to the carrier's representations as to quantity, description and state of the goods. The users also agree that a Bolero signed message shall be admissible before any court or tribunal as evidence of the message. The whole structure is maintained by a 'Trusted Third Party' (TTP) which acts as arbitrator.

The system is essentially governed by two registries: the Core Messaging Platform ('CMP') where users communicate electronically with each other, and the Title Registry ('TR') which keeps a record of all Bolero B/L holders and transfers in ownership, thus regulating the parties' rights and liabilities. The system transfers the message, as well as acting as a database, logging and storing messages for future reference. The integrity and reliability of data transfer is guaranteed by using Bolero itself as a TTP for the transmission of messages.

A Bolero (ie, an electronic) B/L is created by taking the CMP one stage further through to the TR. The registry records the same basic data contained in a paper B/L, including the parties, description of the goods, date of shipment and the terms of the carriage. Authorised parties are entitled to access the TR to determine the status of the B/L at any one time, identify the party 'holding' the B/L and view all prior endorsements made on the electronic B/L, just as they would be able to do with a paper B/L.

A typical transaction starts with the carrier issuing a Bolero B/L to the shipper by transmitting a message to the CMP using a unique digital signature. All messages sent by a user via Bolero must be digitally signed by the user. To sign a message, the user has his own private key, which serves to identify him. The CMP will verify the user's digital signature, log and store the message and then relay it to the recipient. If the user wishes to do so, he can encrypt the message so that it can only be read by the intended recipient.

If the shipper wishes to transfer the Bolero B/L to a third party, such as a buyer or a bank, he will electronically inform the CMP, which will verify the shipper's electronic signature and check the message for authenticity against the TR. If all parameters match, the new owner of the Bolero B/L will be listed into the TR as the last holder in time of that bill with title to the goods. The procedure is repeated for each subsequent operation until delivery, at which time the 'legal owner' of the Bolero B/L will be able to claim the property in the cargo by identifying himself at the docks. The whole sequence of events reflects the stages in the paper-based transaction, except for the addition of the TR, which records each transfer of ownership of the Bolero B/L. This extra element does not infringe the requirement of free negotiability 'without notice'. No human authorisation to the transfer is ever required. The system of validation against the TR of each transferee/endorsee's details represents an invaluable safeguard against the practice of forging documents. The Rule Book clarifies the role of the parties to the B/L. The parties are still designated as shipper or party consignee, for example, but to make the bill of lading transferable, their status is also defined by their holding role. There are four designated categories of 'holder':

- the holder to order;
- the pledgee holder;
- the bearer holder; and
- the consignee holder.

Because physical possession of the B/L represents entitlement to physical delivery, by depositing a B/L with a third party by pledge, a party can raise finance against security for the goods. The pledgee takes security over the goods without becoming liable as a party to the contract. For the first role of the B/L, the physical handling of the paper bill of lading over to the pledgee is replicated in Bolero by the carrier notifying to the pledgee that he now holds the goods to the order of that party.

At English law, it is not possible to replicate the assignment of a B/L by a seller to his buyer in the electronic environment for several reasons. First, section 136 of the Law of Property Act 1925 provides that assignments which are not in writing can only operate by way of equitable assignments and section 25 provides that dispositions of equitable assignments must be in writing and signed by the person disposing of the same. Secondly, CoGSA 1992, which provides the statutory framework for circumventing the problems of privity of contract to enable an assignee to stand in the shoes of the shipper and to sue the carrier for any damage to the goods as if he had all the rights to the original contracting party, expressly states that it does not apply to electronic B/Ls.

For the second role of the B/L, the Rule Book has replaced the concept of assignment by novation so that, instead of the assignee standing in the shoes of the original assignor and suing under the original B/L, the electronic B/L comes to an end and is novated by the shipper and the carrier in favour of the third party buyer on the same terms as the original B/L.

For the third role of the B/L – the document of title – the lawful holder of the e-bill transfers his interest in the goods to his counterparty by the legal mechanism of attornment of the e-bill to the counterparty.

For the oil trader in particular there are a number of advantages with Bolero. It avoids delay in the B/L catching up with the actual trade, it reduces the costs of carrying the bill of lading around the world to the various interested parties, and it should reduce forgery. For carriers, however, the position is somewhat different. They have been using the traditional B/L system for centuries and for them there is no reason to change a system that works. For a carrier, there are limited savings in signing up to Bolero, save for the commercial pressures that are exerted upon them by their counterparties ie, the traders. At present, they simply issue a B/L at the port of loading, carry the cargo to the port of destination, receive the B/L or LoI and discharge the cargo. The carrier remains unconcerned about the transfer of documentation between these two points. A complication is, however, that the carriers' P&I clubs, while giving a cautious welcome to Bolero, have refused to insure them for risks that may arise from the electronic world which would not have arisen in the paper world before passing these risks on instead to the general insurance market.

The Baltic and International Maritime Council (BIMCO) has recently developed and published a charterparty clause that specifically addresses the use of electronic B/L. The increasing use of electronic documentation, particularly in the dry cargo sector, has resulted in a growing user demand from owners and charterers. Today, there are two platforms at the forefront of developments in electronic B/L systems – essDOCS[29] and Bolero – both of which have been approved by the International Group of P&I clubs ('the International Group') as well as by P&I clubs outside the International Group.

In the very recent case of *Glencore v MSC,*[30] one container of cobalt was not delivered and two others were misappropriated, ie, given to another party. When the cargo arrived at Antwerp, it was handled under an electronic release system ('ERS') used for containerised cargo at the port terminal. In essence, under the ERS carriers do not issue paper delivery orders or release notes against B/Ls, but instead provide computer-generated electronic numbers (or 'import pin codes'), which holders of bills present to the terminal and so take delivery of their goods. The B/L in the present case contained this express term:

> *If this is a negotiable (To order/of) Bill of Lading, one original Bill of Lading, duly endorsed must be surrendered by the Merchant to the Carrier...in exchange for the Goods or a Delivery Order.*

Glencore argued that, as a matter of construction of the B/L or by implication, MSC should therefore have delivered the cargo only on presentation of the bill or a delivery order given in exchange for it. The ERS had been introduced with effect from the start of 2011. Under it, when a consignee or its local agents presented a B/L and paid freight and other charges, the carrier sent to it at a designated email address a release note, which provided the addressee with a pin code. Each code was automatically generated by the system and corresponded with a code stored (under encryption) in the Antwerp Port Authority's database. The judge found that the shipping company and its agent had no access to it. The ERS was introduced as an alternative procedure to the existing practice at Antwerp whereby, when B/Ls were

29 Formerly Electronic Shipping Solutions (ESS). See www.essdocs.com.

30 *Glencore International AG v MSC Mediterranean Shipping Co Ltd* [2015] EWHC 1989 Comm.

presented, carriers exchanged them for documents that were presented to the terminal to take possession of goods. It was thought carriers would see commercial advantage in adopting the more efficient ERS, but it was not mandatory and, even today, it has only been adopted by some of the carriers who operate to the Port of Antwerp. Between January 2011 and June 2012, Glencore made 69 shipments of cobalt (approaching an average of one per week) that were carried by MSC to Antwerp under B/Ls with terms materially similar to those of the B/L in the present case. On each occasion, Steinweg acted as Glencore's agent to present the original B/L to MSC or MSC Belgium and, on each occasion Steinweg used a pin code to take delivery of the goods from the 'MSC Terminal'. MSC's argument was that it had complied with the express terms of the B/L, in that the electronic pin codes provided to Steinweg constituted a 'Delivery Order' within the meaning of the B/L. The court rejected that argument.

> *As is explained in Benjamin's Sale of Goods (9th ed, 2014) at para 18–212, the expression "delivery order" is used to describe documents of different kinds: it must be interpreted in its context in the B/L. To my mind, the parties must be taken to be referring to what is commonly called a "ship's delivery order", an expression used and defined in [the] Carriage of Goods by Sea Act, 1992 s. 1(4). It is an essential feature of such a delivery order that it contains an undertaking given by the carrier who is party to it (or possibly assumed by the carrier through attornment: see Benjamin, loc cit, at para 18–216) to a person identified in it to deliver the goods to which it relates to that person. This is required by the statutory definition, and is in accordance with usage before the 1992 Act: see Krohn & Co v Thegra NV [1975] 1 Lloyd's Rep 146, 154–155. The usual purposes of agreeing that a delivery order, rather than the goods, may be delivered in exchange for a bill of lading is to expedite the performance of contracts and particularly to allow bulk cargoes to be split into parcels, without resorting to the practice (a practice "fraught with danger", per Longmore J in* Noble Resources Ltd v Cavalier Shipping Corp *(The "Atlas") [1996] 1 Lloyd's Rep 642, 644) of issuing substitute bills. It is readily understandable that a shipper should agree for practical reasons to this much flexibility in performance. It strikes me as improbable that it would agree to a term whereby the holder of the bill of lading might surrender its rights under it against the carrier without receiving in return either the goods themselves or the benefit of a substitute undertaking from the carrier. There is no need to interpret the B/L so as to have this improbable effect.*

The court held that it was not enough for MSC to establish that, in view of the previous pattern of dealings, the B/L allowed it to use the ERS, and so not to deliver the goods or a delivery order immediately the B/L was surrendered but only deliver the goods when the codes were presented to the Port Authority. It needed to show that, when MSC Belgium provided pin codes for previous shipments, it was taken thereby to have fulfilled MSC's responsibilities with regard to the cargo and its delivery, so that MSC was taken to have no further liability. The previous dealings provided no support for this proposition. MSC, or at least MSC Belgium, wanted to adopt the ERS, but, even if it be supposed that Glencore was content for it to be used, it does not follow that it would have agreed that delivery was considered to have been made when it, or Steinweg, was sent the pin codes rather than when the goods were later received. Indeed, the operator's covenant contemplated that delivery took place only when the code was used to take the goods from the Port Authority: it referred to containers being delivered only once that consignee or its representative had entered the container number and corresponding code into the operator's system, and expressly stated that the conditions were "without prejudice to the applicable legal and contractual provisions covering liability for loss or damage to cargo". It went on to provide that:

No right of delivery may be conferred by any codes or references other than the [specified] release code …

The covenant therefore contemplated that receipt of a code conferred a 'right of delivery', rather than constituting delivery itself.

This case is a reminder that electronic documents and pin codes are no substitute for actual delivery.

12.5 The future of paperless global trade

In 2010, the CMI published a set of Rules for Electronic Data Interchange ('EDI Rules'). The EDI Rules apply where the parties to a contract so agree. Article 7 deals with 'Rights of Control and Transfer' (both of which concepts are defined elsewhere), and concludes by stating that "the transfer of the Right of Control and Transfer in the manner described above shall have the same effects as the transfer of such rights under a paper bill of lading."

Once established that 'negotiability' is achievable, it remains to be seen to what extent operators across the globe will rely on EDI technology to conduct their day to day business. One legal concern is that, while Bolero allows switching back to a paper B/L whenever one of the parties is unable to continue the EDI transaction, there is no possibility of returning to EDI mode. Similarly, a procedure whereby a transaction which starts off in paper format can be converted mid-voyage is not permitted under the Bolero system. To this extent, the development of an IT global network will be an essential prerequisite to the expansion of Bolero or like systems. Furthermore, the approach of legislators will also have an enormous impact on the success of EDI transactions on a global scale.

Nonetheless, EDI and paperless B/Ls or electronic bills are a feature of the industry that is here to stay and is prevalent in the tanker industry in particular.

13. Stale and spent bills of lading

A final word must address what is meant by these terms. In a leading case,[31] the issue concerned was whether a B/L was stale, having been issued on April 6, whereas loading of the cargo was completed on or before March 24. The judge said:

At first I thought that this was a good point, because this is prima facie *a long interval of time and 13 days' delay was held to be fatal in Hansson's case (Hansson v Hamel & Horley Ltd [1922] AC 36, [1922] All ER Rep 237). However, as counsel for the sellers pointed out, in that case the bill of lading was issued not only after the ship had sailed,*

---

31 *M Golodetz & Co Inc v Czarnikow-Rionda Co Inc, The Galatia* [1979] 2 All ER 726, per Donaldson J at 740–741.

but after it had arrived in another country. There is no obligation to issue bills of lading hold by hold. All that is required is that they should be issued within a reasonable time of completing loading. Here the intention was that there should be a single bill of lading covering the whole parcel. This would have been issued on or about 6 April when loading was completed. In the event, the bill of lading was split and both bills were issued on this date. The buyers did not object to the other bill of lading as being stale and I do not think that they can object to this one on this ground. It was issued as soon as reasonably practicable after the completion of the loading of the whole parcel and at or about the time when the ship sailed.

As for the expression 'Spent B/L', assistance can be found from a case[32] in which the judge observed:

I think the bill of lading remains in force at least so long as complete delivery of possession of the goods has not been made to some person having a right to claim under it.

In another case,[33] the question of the scope of the phrase "at a time when possession of the bill no longer gave a right (as against the carrier) to possession of the goods to which the bill relates" is less easy.

[Counsel for the shipowners] submits that the phrase deals only with the case of "spent" bills of lading, ie bills which cover goods that have been delivered to the person entitled to delivery under the bills. [Counsel for the shipowners] submits that the wording was not intended to deal with the situation where the goods never arrive at the destination under the contract of carriage because they have been destroyed or totally lost. But [Counsel for the FOB buyer] points out that in the Explanatory Notes to the draft Carriage of Goods by Sea Bill which is annexed to the Law Commission Report, it gives an example of what is intended to be covered by the wording of clause 2(2) which became section 2(2). The note says: "The words 'possession of the bill no longer gives a right to possession of the goods' cover inter alia the case where delivery of the goods has been made and also the case where the goods are destroyed.

It is clear from the wording of section 5(4) of the Act that the draftsman had well in mind the possibility that a bill of lading could be issued and then the goods covered by it "ceased to exist". The Explanatory Notes to the Bill state that this subsection "makes it clear that rights of suit in relation to any document can exist in respect of goods carried on a vessel which sinks". Therefore there is no reason to confine the circumstances covered by the words "possession of the bill no longer gives a right (as against the carrier) to possession of the goods to which the bill relates". How the Act operates depends on its wording. That is emphasised by the particular reference to sections 2(2) and 4 in section 5(4).

In my view, valuable assistance on the ambit of the words in section 5(2)(c) "at a time when possession of the bill no longer gave a right (as against the carrier) to possession of the goods" is gained from the analysis of Lord Hobhouse in his speech in The Berge Sisar. In para 31 of his opinion Lord Hobhouse notes that the 1992 Act is concerned solely with contractual obligations created in a bill of lading in relation to the

---

32 *Glyn Mills Currie & Co v East & West India Dock Co* (1882) 7 App Cas 600, per Willes J.

33 *The Ythan* [2006] 1 Lloyd's Rep 456, per Aikens J at 475–476.

carriage and delivery up of the goods. He emphasises that the Act is not dealing with proprietary rights of anyone who becomes a holder of the bill of lading. This distinction is important. It means that when a ship sinks and the cargo carried under a bill of lading is lost permanently, the question to ask in connection with the wording in section 5(2)(c) under consideration is: does possession of the bill of lading any longer give a contractual right (as against the carrier) to possession of the goods to which the bill relates? Like the authors of Carver, my view is that there cannot be a contractual right (as against the carrier) to possession of goods that no longer exist (for practical purposes) because they are at the bottom of the sea. If the reason for the loss is a breach of contract by the carrier, there may at that stage spring up a contractual right to damages, but whether there is and who can exercise that right are different questions which I need not discuss here.

So I conclude that the phrase in section 5(2)(c) "at a time when possession of the bill no longer gave a right (as against the carrier) to possession of the goods to which the bill relates" do apply to a situation where the goods have been lost forever, as in this case. That does not stop the Act operating to transfer rights of suit or liabilities. Whether that occurs depends on whether the conditions set out in the Act, particularly those in sections 5(2), 2(2) and 3(1) are fulfilled.

14. Conclusion

A bill of lading is a document signed by the shipowner, or by the master or other agent of the shipowner, which states that certain specified goods have been shipped in a particular ship. It sounds simple. As the preceding discussion has attempted to demonstrate, however, the B/L is a unique document which evolved out of necessity to provide practical assistance to merchants involved in the buying, selling and transportation of all types of goods. It has evolved unique features, such as the fact that it is a negotiable instrument and a document of title, and can justly be said to be a unique creature and a testament to judicial and common sense. It is continuing to evolve and we see that in the form of the electronic B/L. It is also a creature of the common law system, without which it could never have evolved and, in the process, has provided joy, confusion, angst, pleasure and reward to lawyers, judges, academics and law students alike for several centuries past and hopefully a few to come.

Trade finance

Simon Cook
Anna Koshy
Sullivan & Worcester UK LLP

1. Introduction

The term 'trade finance' is used generically to cover a variety of methods and structures for financing international trade flows. Trade finance is typically used to fill funding gaps that may arise at various stages in the supply chain process, from producing and manufacturing goods, to selling them to buyers across the globe. Trade finance instruments and financing structures tend to focus on the mechanics of specific movements of goods and payments between parties. Consequently, providers of trade finance need to understand closely the particular trade process that is being funded, in order to identify and minimise the key risks.

The most suitable funding structure will invariably depend on several factors, including who the parties are, what goods are being financed and where the goods are located. The aim of a good financing structure is to control the flow of commodity and money so that the borrower gets access to funds at the appropriate time and the financier is reassured, so far as possible, that the risks of non-payment by the borrower have been mitigated. This level of structuring and control means that default rates on trade finance transactions have historically tended to be low.

This chapter discusses some of the financing structures commonly used for oil and gas transactions, including the forms of security that are usually taken. It also considers the legal nature of some of the risks frequently associated with dealing in oil and gas, such as environmental considerations and regulatory issues. The impact of English law on trade finance issues is also considered. Because of the nature of oil and gas transactions, however, there are likely to be relevant legal considerations in more than one jurisdiction. In addition, the key issues are analysed primarily from the point of view of a financier, though many of the points raised will also be relevant to commercial parties involved in the buying, selling and distribution of oil and gas.

2. Common financing structures

This section discusses common 'structured' financing methods for oil and gas. For a financier, one of the most important aspects of structuring a transaction properly is to ensure that the risks of a borrower or counterparty defaulting on a transaction are mitigated or controlled, so far as possible. While the appropriate financing structure will be guided by the borrower's requirements some of the key questions to consider are as follows.

- For what stage of its business does the borrower require funding? For example, if the borrower needs money to produce the goods themselves, before they are sold, it is likely that the most suitable structure will be a pre-export financing. This is discussed in more detail below.

- How will the loan be repaid? The financier needs to understand and manage the payment flows that will be used to repay the loan. A fundamental aspect of trade finance is that it is self-liquidating. This means that the revenue generated by the underlying transaction (eg, the production and sale of crude oil) is segregated from the rest of the borrower's business and can be applied in repaying the funds that were borrowed. This is in contrast to certain other forms of lending, which may rely on the success of the borrower's business as whole, such as balance sheet lending.
- What is the commercial process for dealing with the goods from production through to delivery and purchase? This mainly affects what type of security can be taken. It is not usual in trade finance to take a floating charge over the whole of the borrower's assets, because the funding is typically targeted to a specific transaction or flow of goods. As such, the financier will usually look to take control of the goods and eventually the payment flow from the sale of those goods by way of the use of third parties (eg, a collateral manager), restrictions and undertakings in the documents and the taking of appropriate security at the relevant stages of the transaction. Other matters relating to taking security will also need to be considered, such as whether the security needs to be registered and, if things go wrong, what the process for enforcement of that security will be. The financier will want to be able to liquidate the assets or ring-fence funds and appropriate them as quickly as possible in the event of a serious default by the borrower.
- Who is the borrower? Of course the financier will need to do significant due diligence on its borrower from a financial perspective as well as for regulatory reasons, such as complying with 'know your customer' requirements. The laws of the country in which the borrower is incorporated (or conducts its business) will also have an impact on a number of transaction- specific issues: eg, does the borrower need licences or authorisations in that jurisdiction to carry on its business and does it have the ability to borrow and grant security? Typically, these kinds of 'capacity' points will be covered in a legal opinion from a law firm in the jurisdiction of the borrower.

Some of the more common structures used in the financing of oil and gas are considered below.

2.1 Pre-export and prepayment finance

Under this structure, the borrower will typically be a producer. The borrower will produce the goods and sell them to a buyer (usually in a different jurisdiction). The buyer will purchase the goods from the borrower and make payment into a designated collection account, which will be in the name of the borrower but controlled by the financier. The sale proceeds paid into the collection account will be used to repay the loan.

The documentation for this structure will comprise the following.

- A loan agreement between the financier and the borrower, which would typically include (among other things) the mechanics of drawdown and repayment, representations, warranties and undertakings from the borrower in respect of itself and the relevant transaction being financed, and events of default. It may also include coverage ratios and security 'top up' clauses.
- An assignment, in favour of the financier, of the sale contract between the borrower and its buyer(s). This will give the financier rights over amount(s) due from the buyer(s).
- A collection account security agreement giving the financier rights to the balance in the collection account for repayment of the loan, and restricting access to those amounts by the borrower.
- The financier may also wish to take security over the goods themselves while they are (a) in storage at the location of the borrower, and/or (b) in transit to the buyer. Whether this security is required will largely depend on the costs and practicalities of taking that security, which must balanced against the risks of not taking it, as well as when sale proceeds are payable and title passes to the buyer(s).

In terms of security for the sale of oil and gas in particular, the financier may also look to take control of the bills of lading if the goods are being shipped prior to title passing to the buyer(s). Furthermore, if any letters of credit are provided by a buyer, the financier may wish to take an assignment of the proceeds of that letter of credit, or perhaps even be a nominated bank under those letters of credit. Bills of lading and letters of credit are discussed below in further detail.

In addition, in most structured finance trade transactions, the sales contracts will form a key part of the repayment process, so the financier will need to review the sales contracts and be reassured that the delivery and payment terms under those sales contracts work together with the repayment profile under the loan agreement. The financier will also need to ensure that the rights under the sales contract are assignable (or if necessary, get consent from the debtor for such assignment).

2.2 Prepayment finance

In a prepayment structure, the borrower is usually the offtaker (ie, the buyer) of the goods, but is usually not the end buyer. The offtaker will use the proceeds of the loan to make an advance payment to the producer for goods that the offtaker has agreed to purchase. The producer will use the advance payment to produce goods for sale to the offtaker. The offtaker will then sell those goods onwards to an end buyer, and the proceeds of that on-sale will be used to repay the loan.

The prepayment arrangement between the offtaker and the producer will be documented separately from the loan (either in the sale contract itself or in a prepayment agreement). For oil and gas production, prepayment financing is usually revolving, in order to allow ongoing financing of the production and sale cycles.

The producer may invoice the offtaker for a portion of the price of each shipment made to the offtaker, and deduct a certain amount from each invoice to account for the advance payment. This will ensure that the producer retains some cash flow. Alternatively, the parties may agree a discounting mechanism in the offtake contract so that the producer retains a profit element.

The offtaker will want recourse under the loan agreement to be limited, as its ability to perform is dependent on the producer fulfilling its obligations. The financier will, however, want as much recourse as possible to the offtaker since the latter is the borrower and, if one of the larger traders, also a good credit. On balance, financiers will typically expect an offtaker to share a certain (sometimes limited) amount of the risk on the producer, so that the offtaker has a vested interest in pursuing the producer in the event of a problem arising. If available, this would commonly be in the range of 10% – 20% of the risk. Once the producer has performed and has delivered to the offtaker, it would be usual for the offtaker/borrower to be on full risk for that part of the loan.

As with pre-export financing, the financier needs to be comfortable with the key commercial contracts in the transaction, ie, the prepayment agreement and the sales contracts with the producer, the offtaker and the end buyer(s). If deliveries under the prepayment agreement will be made over a long period, one point for the financier to be aware of is that the price payable by the offtaker may vary in line with market fluctuations. To deal with this, the financier will need to ensure that there is a mechanism in the prepayment agreement for the value of deliveries to match the advance payment amount, and a corresponding obligation on the borrower to pay if there is a shortfall.

On the security side, the financier will seek to take assignments of the key rights of the offtaker under the prepayment agreement and the sales contracts with both the producer and the end buyer(s). This may include an assignment from the offtaker of the right to take delivery under the offtaker sale contract. There is also likely to be a collection account for end buyer payments to be collected and security over the goods when they are owned by the offtaker and, in some cases, the producer as well.

2.3 Borrowing base facilities

Borrowing base facilities are usually revolving and can be particularly suitable for oil and gas producers who will have goods in various stages of production over which they can give security, including commodity still in the ground. In short, the amount which a borrower can draw under a borrowing base facility at any time will depend on (and be a specified percentage of) the value of the secured assets at that time. There is usually a coverage ratio that must be maintained by the borrower which protects the lender against fluctuations in the value of the security while the financier has exposure under the facility. Under the coverage ratio mechanism, the borrower is typically required to pay cash into a blocked account or provide further security if the coverage ratio is not met.

The financier will need to monitor closely the assets being secured. This can include having to ascertain actual and perhaps even predicted levels of production, and the quantity and quality of goods in storage and reserves. The financier may also seek to take security over other relevant assets of the borrower, such as the production licence and key equipment and contracts.

The financier will need a method for valuing commodity at different stages of production, as there will be different risks at each stage. For example, in relation to unextracted commodity, the value of that commodity will need to take into consideration whether the predicted reserves will actually yield the expected amount of commodity. For the purposes of determining the maximum amount that can be drawn at any given time, the financier will apply different discounts for the different stages of production, from commodity in the ground to cash in a segregated account from commodity which has been sold. The discount will be greater for an asset which is harder to liquidate (eg, unextracted commodity).

The type of security taken over the goods will depend on where the goods are located and what state they are in. A floating charge over the secured goods may be taken, or a pledge over identifiable lots of goods from time to time, if practical. There will need to be a balance between, on the one hand, the financier's requirement to have sufficient goods to meet outstanding payment obligations and, on the other, the borrower's requirement to run its business without too much interference from restrictions created by the security.

3. Security and control of commodities

As part of the financing structures previously noted, the financier will seek to take security over the relevant goods, documents and/or payment flows related to the transaction. The type of security taken will depend on a number of factors, including the nature of the commodity itself, the location of the commodity and whether registration is required or stamp duty is payable. Security will normally be specific to certain assets or documents and not a general 'fixed and floating' type security over all the borrower's assets that is more usual in working capital facilities.

The key elements of the most common English law security taken in oil and gas financing structures are summarised below.

3.1 Pledge and trust receipts

In most jurisdictions, a pledge usually has one distinctive character – it requires the financier to have actual or constructive possession of the pledged assets in order to be effective. Actual possession means the financier physically controls the asset, whereas constructive possession means a third party, such as a collateral manager, holds the assets on behalf of the financier. The latter is more common in financing situations in which the financier may not be in the location of the goods in order to possess them physically, or may not have the resources to possess them. Actual possession under a pledge agreement is frequently seen in relation to letters of credit where a financier is involved in the letter of credit (LC) process. A pledge would commonly be taken over title documents relating to the goods which pass through the counters of the financier when presentations are made under the LC. This means that the financier will have control over the title documents at that time and therefore over the goods they represent.

Under English law, there is a distinction between possession and control of the goods. While only possession is required for a pledge, the distinction is not always easy to determine. There is no requirement to register a pledge in England & Wales, which makes it an attractive option for security, but a financier would normally take a pledge under the law of the jurisdiction of the location of the goods. For goods moving cross-border, eg, through pipelines that cross several countries, taking pledges in each jurisdiction that the goods enter is likely to be impractical, so a pledge would not be suitable. A financier may, instead, seek to take security over the borrower's rights under the pipeline agreement under which the oil or gas is transported.

The borrower will need to take possession of the goods at some point in the transaction in order to sell them and generate proceeds to repay the financier. At such time, the financier will have to lose possession of the goods, so the pledge will no longer be effective. In order to allow the borrower's business to function and the goods to be released for sale, under English law (or that of any common law jurisdiction which recognises a trust), the financier can obtain a trust receipt from the borrower in return for releasing the goods. Under this document, the borrower agrees that, following release of the goods by the financier, the borrower will only use them for the agreed purpose (ie, sale to the end buyer) and will hold the goods, and consequently the proceeds for sale of the goods, on trust for the financier during the period the borrower has possession. This gives the financier control over the goods and cash flow in the gap between the pledge being released and money being received for the goods. The goods and related sale proceeds will be held on trust and would therefore fall outside the borrower's estate on insolvency.

3.2 Assignment

An assignment of receivables generated by sale proceeds and/or certain contractual rights under a sale contract (eg, to accept delivery of goods) is usually an important part of the security package. An assignment gives the financier ownership rights over the assigned assets, with a corresponding right for ownership to be returned to the borrower when the secured obligations are satisfied.

Under English law, there is a distinction between a 'legal assignment' and an 'equitable assignment'. A legal assignment must (among other things) be in writing and the debtor must be given notice in writing of the assignment. An acknowledgement from the debtor of this notice is not legally necessary under English law (though it may be required under the laws of the debtor's jurisdiction). It is usually useful to get an acknowledgement in any case as it serves as good evidence that notice was served. Multiple notices served on a debtor for the same debt would take effect in priority from the earliest notice served. Once notified of an assignment, the debtor would have to pay the assigned proceeds as directed under the notice of assignment, otherwise they would be deemed not to have discharged their debt.

Under a legal assignment, the financier would be able to take action directly against the debtor if necessary to enforce its rights under the assignment. If the requirements of a legal assignment are not met (eg, it is not in writing or notice is not served on the debtor), an equitable assignment would be created. This would still give the financier certain security rights but they are generally considered to be less robust than rights under a legal assignment and would require the assignor to be joined in any proceedings against the debtor. Equitable assignments are, however, still a useful, and sometimes unavoidable, form of security.

Another point that is typically included in assignment agreements for structured financing is a continuing requirement on the borrower to fulfil its obligations under the relevant sale contract and, correspondingly, the rights of the financier not to have any involvement in the underlying transaction. In some cases, however, a financier may wish to have step-in rights so that it can take delivery and/or sell goods to the buyer on behalf of the borrower if the borrower fails to do so itself.

3.3 Charge

A charge grants a financier a security interest over an asset. The financier would have the right to appropriate the charged assets and use them in satisfaction of the relevant debt (or, if the charged asset is commodity, to sell it and use the proceeds in satisfaction of the debt). A charge can generally be fixed or floating, depending on the degree of control that the financier has over the asset.

In a trade finance structure, a fixed charge would usually be taken over a collection account and the financier would have control over withdrawals from that account by the borrower. The borrower may require access to any excess funds in the account which remain after the financier's due obligations have been paid. A financier should take care not to release such excess funds automatically as this may be determined to be a lack of control, in which case the charge would be deemed to be floating. Under English law, a floating charge would rank after a fixed charge over the same asset (and fixed charges over the borrower's assets generally) on insolvency of the borrower.

If the charge is to be created over a pool of assets that may change from time to time, eg, as goods are sold to buyers, a fixed charge may be impractical as the borrower would need control over those goods to run its business. A floating charge gets around this issue as it essentially 'floats' over the assets and only attaches to the assets available when the charge crystallises (usually upon the occurrence of an event of default).

The appropriate charge (or its equivalent in the relevant jurisdiction) will depend on what the asset in question is and whether the borrower needs access to that asset in order to run its business. In oil and gas transactions, it is less usual to take a fixed charge over the commodity, as the borrower usually needs to be able to use the commodity in its course of business. This means that the financier is unlikely to have the requisite control for a fixed charge.

4. Alternative financing and payment structures

The financing structures identified so far are based around debt financing, ie, lending structures. This section considers some alternative structures used in the oil and gas sector.

4.1 Commodity repurchase structures

One alternative approach to a lending structure is for a trader actually to sell the commodity to a financier and repurchase it at a later date. This has become increasingly common, perhaps due in part to increased regulation for banks around lending.

There are a number of advantages for the seller and the financier to this type of transaction. In particular, if structured correctly, the seller will get off-balance sheet treatment for the sale of commodity to the financier. This can be useful where a company is highly leveraged or a financier does not have the appetite for further lending to a company. The balance sheet treatment achieved under a commodity repurchase (repo) will depend on what accounting standards are applied and whether the transaction is a 'true sale' (as described below).

From the financier's perspective, it is the owner of the goods, not a secured party; therefore, if the seller fails to repurchase the goods, the financier does not have to go through a potentially cumbersome process of enforcing security. Instead, the financier can sell the commodity outright to a third party. Depending on the commodity in question, the financier may require the commodity being purchased to be tradable on an exchange in order to make any on-sale to a third party simpler.

Another scenario in which a repo structure is helpful is where the security a financier would seek to take is simply not feasible in the relevant jurisdiction or where the prerequisites for that security cannot be met (eg, possession for a pledge).

Commodity repo structures would not be a suitable option where pre-export stage or advance payment stage financing is required, as the commodity has to exist, and be owned by the seller, in order for it to be sold to the financier. In certain situations, however (eg, oil being shipped or gas in storage), commodity repos are a useful alternative financing tool.

There are a number of points to consider when entering into a commodity repo. The most important is the risk of recharacterisation. If a commodity repo is not structured properly, there is a risk that the transaction could be construed to be a loan rather than a sale. This is on the basis that (in its simplest form) the payment from the financier to the seller for the commodity is treated as a loan, the transfer of commodity from the seller to the financier is potential security for that loan, and the 'repurchase' by the seller is deemed to be 'repayment for the loan.

Whether a transaction is likely to be recharacterised will depend on the law of the relevant jurisdictions involved, namely:

- the law of the location of the goods;
- the governing law of the sale and purchase contracts; and
- the law of the jurisdiction of the seller.

If a transaction is recharacterised as a loan, the financier is likely to be left as an unsecured creditor in the case of the seller's insolvency. This is because, in most cases, any formalities for creating security (eg, that the security may need to be in writing, or registered) would not have been completed as the parties did not intend to create security. In some cases it may be possible for 'fall back' security to arise automatically without any additional formalities being required. In most cases, however, it would not be wise for a financier to rely on the possibility of a 'fall back' security by complying with the required formalities in parallel, as this means the transaction is more likely to be recharacterised as a loan. In some jurisdictions where the risk of recharacterisation is high (eg, the US), it is advisable to take 'fall back' security expressly. Financiers must, however, be careful with this approach, as taking security may have a negative impact on the recharacterisation risk in other jurisdictions relevant to the transaction.

The best approach is to ensure that the key element of a sale, ie, a transfer of title in the asset, is effected in accordance with the appropriate laws related to the transaction. Ideally, the financier will hold either a title document (if this exists in the relevant jurisdiction for oil and gas) or the best evidence of title, such as a tank certificate, upon a sale of the goods.

For a sale to occur under English law, title and risk in the goods must pass to the buyer and the goods must be delivered to the buyer. For delivery, where the goods are held by a third party, it is important that the third party 'attorns' to the buyer that the goods have been delivered to the buyer.

Many commodity repo transactions in the market are documented under English law because of the relatively favourable treatment of these structures. One of the key principles of English contract law is that the intention of the parties to create certain legal relationships will be respected. Therefore, if a sale contract governed by English law is clearly documented as a sale (with appropriate transfer of title and risk, and delivery obligations) it is likely that an English court would treat this as a sale. In order to evidence the intention to sell and not create a loan arrangement, however, the transaction needs to be viewed as a whole. Some of the key points to consider during the financier's period of ownership are as follows.

- Who takes the risk of damage or loss to the goods?
- Who procures insurance for the goods?
- Who takes the risk of price fluctuations?
- Does the seller have an option or an obligation to repurchase?
- Does the purchase and resale price reflect market value?

The more risks (and benefits) of ownership that are transferred to the financier, and therefore the more a transaction resembles a market-standard sale, the more likely it is that the transaction will be treated as a sale and not recharacterised.

Another aspect of commodity repos for financiers to consider is that they will be an owner of physical goods. In the case of oil and gas in particular, this entails a number of considerations, such as who bears the risk of environmental damage, whether title can pass to a commingled bulk of goods, and whether any licences or authorisations are required for the ownership or sale of goods in the relevant jurisdiction. These points, and others, are discussed in further detail below.

In addition, the financier will own and therefore have to arrange for storage of the goods it has purchased. As many financiers will not want to manage the storage and maintenance of oil and gas (or will not have the resources to do so) a third party will normally be engaged to do this on the financier's behalf. The best case scenario from a true sale perspective is for the financier to engage an independent storage operator or collateral manager directly in order to store the goods. It is, however, also possible for the financier to engage the seller itself, under a 'services agreement', as the latter will already be set up and equipped to perform this task. This role, if it is to be performed by the seller, has to be carefully documented as a separate and independent role and would need to work with the overall structure so as to ensure that it does not interfere with the transfer of title and delivery aspects or otherwise negatively impact upon the aim of the financier becoming owner of the goods.

4.2 Receivables financing

Oil and gas companies will generate receivables from the sale of oil or gas to their customers. These receivables are frequently payable on deferred payment terms which are usually around 30 days, but can be up to 90 or more days, after delivery. The seller's right to be paid (usually called a 'receivable') is valuable and can be used as a source of financing to bridge the gap between delivery by the seller and receipt of payment by the buyer.

There are a number of different structures that can be adopted for receivables financing but, in essence, the key characteristic is that the financier will purchase a receivable at a discount from the seller of goods. The purchase may be undisclosed to the debtor. In such cases, the financier will take a risk on the seller of the receivable collecting amounts due in respect of the receivable and transferring them to the financier. In addition, from an English law perspective, the assignment of the receivable to the financier in an undisclosed purchase by the financier would not be notified to the debtor. As explained above, this means that the financier will risk not having a priority claim.

The purchase of the receivable by the financier may be with or without recourse to the seller. Clearly, having full recourse to the seller is the best position for the financier, but in most cases a seller will only accept limited recourse as it will want to divest itself of any further risk relating to the sale following payment by the financier. The seller will be required to give certain representations and undertakings about the underlying sale, including that it has done all that is required to oblige the debtor to pay. Any breach of these provisions would give the financier recourse to the seller, but otherwise the financier usually takes payment risk on the debtor. The extent of recourse given by the seller will have an effect on the accounting treatment of the discount, ie, whether it will be considered a true sale or not.

4.3 Silent payment guarantees

A financier may be willing to provide a guarantee for payment of a receivable on a 'silent' basis, ie, the party which has the obligation to pay the relevant receivable would be unaware of the existence of the guarantee. This is quite common in relation to individual oil receivables, such as for high value spot deals that are not supported by a letter of credit.

The financier will take a fee for covering specific events of non-payment, such as political or foreign exchange risks. It is important that the rights of the financier are carefully documented so that an assignment in favour of the financier is effected (and notified to the debtor) if the financier has to pay under the guarantee.

4.4 Letters of credit

Letters of credit (LCs) are commonly used in sales of oil. In its most simple terms, an LC is an undertaking from a bank to pay the beneficiary of that LC if the latter presents the correct documents to the bank (as set out in the LC itself). The effect of an LC is to transfer the payment obligation from the buyer to a bank. Thus, LCs are typically used by sellers who are uncertain about the creditworthiness of a buyer and/or where the buyer and seller are in different jurisdictions. The issuing bank, which is ultimately liable to pay under an LC, will have a separate arrangement with the buyer for reimbursement of any amounts the bank pays out. An example of documents typically required for presentation under an oil sale LC are:

- bills of lading;
- a commercial invoice; and
- a quality and quantity certificate.

LCs have been used in trade finance for centuries and are therefore very well established, and there is much that can be said about their nature and use. An in-depth analysis of LCs is beyond the scope of this chapter. Their key features are, however, as follows.

- An LC is issued by a bank (the issuing bank) at the request of an applicant (usually a buyer), and made in favour of the beneficiary (usually a seller).
- It is (absent fraud) an irrevocable obligation of the issuing bank, otherwise it would not be of much use to the seller in transferring the payment risk.
- It is autonomous, so that any bank which has an obligation under the relevant LC is not required to review the underlying transaction but can accept documents provided at face value, provided that they exactly match the requirement of the LC and do not evidence fraud on their face.
- There are many roles that can be included for banks in an LC structure. There is always an issuing bank, which issues the LC. In addition, there can be an advising bank (which advises the beneficiary that an LC is in place but is not liable to make any payments under it), a confirming bank (which undertakes to pay the beneficiary on presentation of the correct documents and is then reimbursed by the issuing bank), and a nominated bank (a general term for any bank through which an LC is available).
- Nearly all LCs are governed by the Uniform Customs and Practice for Documentary Credits (the current version being known as UCP 600) which is published by the International Chamber of Commerce. This is a set of rules which govern (among other things) the relationship between the various banks in the LC process, the obligations of those banks and certain mechanical procedures, such as how amendments are made and how to treat discrepant documents provided by a beneficiary.
- There are many variations for structuring LCs: eg, they can be payable at sight or at a future date (the latter are known as deferred LCs). Deferred LCs can be discounted by a bank which will pay the beneficiary before the payment date under the LC, on condition that compliant documents to satisfy the LC requirements are provided. The discounting bank will then present the relevant documents to the issuing bank for payment at maturity.

5. Specific issues to consider in oil and gas transactions

This section discusses some of the issues that typically arise in relation oil and gas transactions. They are general points that are likely to be relevant whatever the form of financing or security adopted by the parties.

5.1 Bills of lading and letters of indemnity

Cargoes of oil and gas are frequently transported internationally by ship. For such shipments, the carrier of the vessel will issue a bill of lading to the seller once the goods are loaded onto the ship. The bill of lading has different functions: it will act as:

- a receipt that the goods have been loaded onto the vessel;
- evidence of the contract of carriage; and
- a document of title to the goods.

It is therefore a key document in the shipping process. The bill of lading will be sent by the seller to the buyer, who will need to present it to the carrier in order to prove that the buyer is entitled to take delivery of the cargo at the port of discharge.

In some cases, the bill of lading will not reach the buyer before the shipment has arrived at the relevant port (eg, because there have been back-to-back sales of the oil on ship and the end buyer is not the same as the original buyer with whom the seller transacted). The carrier will not want to discharge the cargo without presentation of the bill of lading from the person claiming to be entitled to the cargo. This is because the carrier would be liable to the rightful party for the value of that cargo if it were discharged to the wrong party. If, however, the buyer waited until the bill of lading arrived to the relevant party, demurrage charges would arise. One solution to this problem is for the carrier to agree to release the cargo against a letter of indemnity instead of a bill of lading.

The letter of indemnity would typically be issued by a buyer and would set out the buyer's obligation to indemnify the carrier if any claims are made by a third party against the carrier for discharging the cargo to the buyer (and not against a bill of lading). Such letters of indemnity are frequently countersigned by a bank, as the carrier may not have a relationship with the buyer. Alternatively, a letter of indemnity can be issued by the seller to the carrier, as these parties will have an existing relationship. The seller would agree to indemnify the carrier for any third-party claims against the carrier for release against its buyer. A seller would usually only issue a letter of indemnity where there have not been back-to-back sales and the buyer at the port of discharge is the same buyer with whom the seller originally transacted. The issuer (or guarantor) of a letter of indemnity should be aware of the wide liabilities that are associated with issuing this instrument.

5.2 Environmental concerns

Financing oil and gas raises a number of issues related to environmental risks. Consequences for breaching environmental obligations can both be costly and include significant fines and/or reputational damage.

Under a loan agreement, a financier will typically include undertakings and representations in respect of borrower compliance with all relevant environmental laws and regulations, and maintenance of relevant consents or authorisations, supported by appropriate indemnities. In this way, a financier seeks protection in the event that a third party brings a claim against it for environmental damage as a 'shadow' director or financier of the goods (since banks tend to have deeper pockets than most corporates) and are susceptible to such claims, whether valid or otherwise. It is also usual for a financier to be loss payee on the relevant insurances of the borrower (for environmental and other related risks). Financiers will need to review the relevant policies and ensure that the appropriate risks are covered and that the insurer has agreed to note the financier as loss payee.

Some financiers will require the borrower to comply with the Equator Principles.[1] These are a framework for identifying and managing environmental risks associated with project financing and loan transactions. They are voluntary but have been adopted by a range of financial institutions, export credit agencies and development agencies and have become a common benchmark in these types of transaction.

If the financier will be an owner of oil or gas under a commodity repo, it will need to consider what steps it needs to take to protect itself against liability from environmental claims. Depending on the jurisdiction in question, liability for environmental damage may apply only to the entity storing the goods, or it may apply to owners who could have strict liability even if they are not in control of the goods at the time the damage occurs.

5.3 Insurance

Insurance is generally an important risk mitigant in structuring oil and gas financing transactions. Insurance coverage can range from environmental risks, to marine and transport cover, to goods in storage. How a financier takes advantage of the insurance protection a borrower has (or may be required to procure) will vary depending on the transaction being financed.

It is usual in loan arrangements for the financier to be named as loss payee on the borrower's insurance. A loss payee will have the right to be paid proceeds of a claim; it cannot, however, make a claim or otherwise enforce the terms of the insurance policy itself. As such, a loss payee is dependent on the acts and omissions of the insured party (the borrower), who may fail to maintain the insurance properly and therefore void the policy. To mitigate these risks, most loan agreements include covenants for the borrower to meet all insurance premium payments, abide by the terms of the insurance policy and provide the financier with evidence that these obligations have been met.

The proceeds of an insurance policy can also be assigned to a financier but, under English law, an assignment of future rights would only take effect in equity. The financier would therefore not have a direct right to claim under the policy itself; it would, instead, have to join with the insured party in making a claim.

It may be possible for the financier to be co-insured on the seller's existing policy for the goods. This would give the financier the right to enforce the insurance contract itself, but it would also give rise to certain duties on the financier, eg a duty of disclosure in relation to the information related to the insured transaction. As the financier is unlikely to have first hand knowledge of such information, it must

1 www.equator-principles.com.

ensure that such obligations and, commonly, any premium payment obligations are expressly allocated to the borrower.

5.4 Title documents

Where the commodity is stored in a storage facility, the owner or operator of that facility will usually issue a document to the person storing the goods setting out what goods have been delivered to the facility and by whom. These documents have a variety of names, including warehouse receipts, holding certificates and warehouse warrants. Depending on the jurisdiction in question, these documents may be 'title documents', ie, they represent ownership of the underlying goods. Where documents are not treated as title documents, they will only serve as evidence of ownership or (more likely) deposit, but will not usually have rights attached to them. For example, they may or may not entitle the benefiting party to take redelivery of the goods from the storage facility.

5.5 Shipper and storage operator liens

The nature of how oil and gas are dealt with means that these commodities are frequently under the control of a third party, such as a tank operator, a shipper or a pipeline operator. These third parties will invariably have liens that may arise either by law or expressly in the relevant agreement with that third party. A lien will usually allow the third party to retain the goods for which it is responsible until all the costs of providing its services have been settled. A lien does not always have a power of sale attached to it, but such a right may be included in the relevant operator agreement or, in some jurisdictions, be applied for through the courts.

A financier can protect itself against the prejudicial impact of the exercise of a lien by retaining the right to step in and pay such costs on behalf of the relevant borrower, coupled with an indemnity from the borrower to reimburse those costs. Although not an ideal scenario, it would allow the financier to discharge any liens quickly – eg, if it needs to enforce any security over those goods. The financier may even be able to agree with the borrower that storage or transportation fees will be settled in advance, or from the proceeds of a loan, in order to try and avoid a lien arising. In some cases, where the financer has a direct relationship with a third party, the latter may agree to waive its rights to any liens.

5.6 Regulatory issues

Regulation affecting financial institutions is continually evolving and responding to market forces and economic pressures. As trade finance is by its nature international, there will be a number of regulatory regimes that can be relevant to a transaction. Trade finance practitioners should be aware that regulation may apply both to financiers and traders. There are a number of regulatory issues to consider in financing transactions generally, such as sanctions, anti-money laundering and bribery concerns, as well as exchange controls. There are also additional considerations concerned with dealing in oil and gas specifically, such as export controls and strategic oil reserves. It is beyond the scope of this chapter to discuss these points in detail, but strategic oil reserves are an interesting point to note.

Strategic oil reserves (also known by other terms, such as 'oil stocking obligations') are reserves that a government may hold (or may require suppliers in its jurisdiction to hold) for emergency situations.

In the UK, 'substantial suppliers' of crude liquid petroleum or petroleum products are required to hold minimum levels of oil stocks. If there is an emergency, those suppliers would have to release the compulsory stock on demand by the government. Under the Oil Stocking Order 2012[2] a "substantial supplier", is (widely) defined as an entity whose supply exceeds 50,000 tonnes of crude liquid petroleum or petroleum products during the relevant period. There are similar rules in many other jurisdictions.

In this regard, financiers will need to ensure that any stocks of oil that they are financing will not be subject to such regulation. These requirements are also relevant for commodity repo transactions in which the financier will be an owner of oil stock. The financier will need to ensure that it does not fall within the scope of any strategic oil reserve obligations in the relevant jurisdiction as an owner. In addition, where there is a third-party storage operator, the financier must review the storage agreement in order to ensure that the former does not have any rights to purchase stocks it is holding to meet government demands.

5.7 Commingling

During storage and transportation, many oil and gas stocks are commingled with goods belonging to other parties – eg, in multi-shipper pipelines or centralised storage tanks. This type of commingling is standard practice in many jurisdictions and it is sometimes impossible to avoid on a practical level. The operator of the storage facility or pipeline will usually have a detailed procedure for recording and identifying which owners have rights to what percentage of the oil or gas in storage. Rights to commingled goods need, however, to be considered under the laws of the jurisdiction in which the goods are stored or being transported, and understood in case the allocation procedure fails or there are insufficient goods with which to satisfy multiple claims from owners or creditors.

As a separate issue, it may also be necessary for the borrower's business that the goods being financed or sold under a commodity repo are combined with other goods in order to create a new product for on-sale to end buyers. A financer will need to consider what rights it can take over goods at the various stages of production and, perhaps more accurately, what rights a borrower has to give to the financier at these different stages.

A financier will need to take into account:

- whether security can be taken over commingled goods in the relevant jurisdiction;
- what rights it will have to commingled goods on enforcement of security if the bulk of goods available is insufficient to meet all owners' or creditors' claims; and
- whether title can pass or security can be given to a portion of commodity in a commingled state.

---

2 2012 SI No 2862. The Order implements EU Council Directive 2009/119/EC.

Treatment of commingled goods will vary depending on the jurisdiction in question. Some useful points to note under English law are as follows.

- Section 20A of the Sale of Goods Act 1979 as amended allows an undivided share in goods forming part of a bulk to be transferred from one party to another if those goods are identified in a contract between the parties and the buyer has paid the price for some or all of those goods. This is useful in a commodity repo scenario where an owner of goods in a bulk wishes to transfer title. It is also useful to loan providers who wish to take security over part of a commingled mass. In other jurisdictions, however, goods being sold need to be segregated before title can pass.
- Where goods of the same specification are mixed, title to part of the mixed bulk will remain with each party who held title before the goods were mixed. This is useful for a financier taking security over goods that will be mixed at some point in time, such as in a pipeline.
- Where goods are substantially changed in nature when they are mixed, eg, when crude oil is blended to create a new product, title to the new product will be delivered to the buyer when it is created, unless the parties expressly agree otherwise. This means that the seller no longer owns the goods and can therefore not sell them, nor give security over them.

There are also likely to be situations in which the rights of each party are not clear cut – eg, if off-specification gas is accidentally introduced into a storage tank. In such case, a question arises over whether each supplier with title to gas in the tank still retains its title. Advice should be taken in respect of the law of the relevant jurisdiction as to what rights a borrower, seller and financier would have in specific situations.

6. Conclusion

Trade finance is a useful tool to assist the production, transportation and sale of oil and gas by providing targeted financing at appropriate stages of the supply chain. Creating a proper financing structure will require a certain amount of due diligence and care, so that each step of the financing and repayment is controlled and, if necessary, secured. Given, however, the relative certainty of financier's rights from the long history of the use of trade finance, and the flexibility of potential options for financing, it is an important part of oil and gas trading.

Oil cargo insurance

Cheryl Guan
John Simpson
Ince & Co LLP, Singapore

1. Introduction

Oil insurance was developed as a sector-focused insurance to support the volatile oil industry and cater to the risks emanating from its activities. Cargo insurance is by far the most frequently purchased insurance by the oil trading sector and is the focus of discussion in this chapter.

The cargo insurance needs of the various players in the oil trading sector are varied. Oil majors benefit from economies of scale and generally self-insure, most commonly through the setting up of related captives. Some oil traders have chosen to band together to form mutual insurance companies, of which Oil Insurance Limited (OIL) is a prominent example. Formed in 1971 with only 16 members, the Bermuda-based OIL today has more than 50 members and insures over $2 trillion worth of global assets.[1]

This chapter is intended to demystify oil cargo insurance for the independent oil trader, who may not have access to captives or mutual insurance companies and must look instead to the general insurance market to meet its oil cargo insurance needs. The discussion will be broad based and will cover the central tenets of oil cargo insurance (including a discussion on the party who is obliged to procure such insurance); the types of cargo insurance coverage available commercially; and some of the common coverage issues encountered in oil cargo insurance.

2. The obligation to procure cargo cover

International trade is the lifeblood of shipping. A cargo can change hands several times before arriving at the intended destination and, as a result, the obligation to procure cargo insurance could easily slip between the cracks in a fast moving trade environment. To clarify the situation and provide a basic contractual framework, the International Chamber of Commerce publishes a series of predefined commercial terms known as Incoterms® Rules or International Commercial Terms ('Incoterms') which set out, among other things, the entity who has the obligation to insure the oil cargo under each particular Incoterm.

There are presently 13 or 11 Incoterm rules, depending on whether Incoterms® 2000 (13 rules) or Incoterms® 2010 (11 rules) are used. Each rule refers to a type of agreement for the purchase and shipping of goods internationally. The table below is extracted from Incoterms® 2000 and Incoterms® 2010 and sets out the classification of the rules in place:

1 See generally Oil Insurance Limited website www.oil.bm (accessed February 24 2016).

Incoterms 2010[2]	Whose obligation to insure cargo
Ex Works (EXW)	No obligation on the part of either party to obtain insurance. The seller has no obligation to the buyer to make a contract of insurance. However, the seller must provide the buyer, at the buyer's request, risk and expense (if any), with information that the buyer needs for obtaining insurance, and vice versa.[3]
Free Carrier (FCA)	No obligation on the part of either party to obtain insurance.
Free Alongside Ship (FAS)	No obligation on the part of either party to obtain insurance.
Free On Board (FOB)	No obligation on the part of either party to obtain insurance.
Cost and Freight (CFR)	No obligation on the part of either party to obtain insurance.
Cost, Insurance and Freight (CIF)	The seller must obtain at its own expense cargo insurance complying at least with the minimum cover as provided by Clauses (C) of the Institute Cargo Clauses (LMA/IUA)[4] or any similar clauses. The insurance shall be contracted with underwriters or an insurance company of good repute and entitle the buyer, or any other person having an insurable interest in the goods, to claim directly from the insurer.[5] When required by the buyer, the seller shall, subject to the buyer providing any necessary information requested

continued on next page

2 Incoterms 2010 came into effect on January 1 2010; see generally International Chamber of Commerce website www.iccwbo.org/products-and-services/trade-facilitation/incoterms-2010/the-incoterms-rules/ (accessed February 24 2016).

3 This applies to all terms where there is no obligation for either party to obtain insurance.

4 These initialisms refer to, respectively, Lloyd's Market Association and Insurance Underwriting Association of London.

5 While there may be an obligation to procure cargo insurance that matches the contract of carriage under which the goods are shipped (see *Belgian Grain & Product Co Ltd v Cox & Co (France) Ltd* (1919) 1 Ll L R 256 and Michael Bridge, *Benjamin's Sale of Goods* (Sweet & Maxwell, 9th Ed, 2014) at [19-049]), any such obligation will be displaced by the express terms of the sale contract. This was considered by the Court of Appeal in *Geofizika DD v MMB International Ltd and another* [2010] EWCA Civ 459 ('*Geofizkia*') in the context of a CIP sale subject to the Incoterms 2000, which require the seller to procure cargo insurance on minimum Institute Cargo Clauses (C) terms, a minimum requirement that was upheld. Although *Geofizika* is a case which turned on its own peculiar facts, it is submitted that it is good law in relation to insurance obligations under Incoterms. Traders who require the more comprehensive cover provided under Institute Cargo Clauses (A) terms would therefore be well advised to provide for this in their contracts.

Incoterms 2010[2]	Whose obligation to insure cargo
	by the seller, provide at the buyer's expense any additional cover, if procurable, eg, cover as provided by Clauses (A) or (B) of the Institute Cargo Clauses (LMA/IUA) or any similar clauses, and/or cover complying with the Institute War Clauses and/or Institute Strikes Clauses (LMA/IUA) or any similar clauses. The insurance shall cover, at a minimum, the price provided in the contract plus 10% (ie, 110%) and shall be in the currency of the contract. The insurance shall cover the goods from the point of delivery to at least the named place of destination. The seller must provide the buyer with the insurance policy or other evidence of insurance cover. Moreover, the seller must provide the buyer, at the buyer's request, risk, and expense (if any), with information that the buyer needs to procure any additional insurance.
Carriage Paid To (CPT)	No obligation on the part of either party to obtain insurance.
Carriage and Insurance Paid To (CIP)	Seller is responsible for arranging insurance cover for the goods. The seller also contracts for insurance cover against the buyer's risk of loss of or damage to the goods during the carriage. The buyer should note that under CIP the seller is required to obtain insurance only on minimum cover. Should the buyer wish to have more insurance protection, it will need either to agree as much expressly with the seller or to make its own extra insurance arrangements. (Refer also to the requirements for CIF, above)
Delivered At Terminal (DAT)	No obligation on the part of either party to obtain insurance.
Delivered At Place (DAP)	No obligation on the part of either party to obtain insurance.
Delivered Duty Paid (DDP)	No obligation on the part of either party to obtain insurance.

3. Overview of the types of oil cargo cover

The discussion below provides an overview of the coverage presently available in the commercial market for oil cargoes. This can be divided into three specific sections, namely:

- loss of or damage to the cargo carried representing a first party loss to the cargo interest;
- third-party liabilities arising as a result of the carriage of the oil cargo; and
- liability arising out of piracy situations where cover may not be afforded under traditional first party cargo cover and third-party liabilities cover.

As the following discussion will show, the cover for third-party liabilities is intended to insure risks faced by a shipowner or charterer in the course of the operation of their business. It is generally the case that traders spot charter vessels and are therefore not normally exposed to third party liability claims from owners, save (for example) claims that they have nominated an unsafe port or berth. For completeness, however, third-party liabilities are covered in the discussion. It is hoped that this will present a full picture of the types of cargo-related insurance products available, even if some classes of cover will only occasionally apply to traders.

3.1 Marine cargo insurance

Organisations such as the Institute of London Underwriters ('ILU') have produced terms for cargo insurance policies which have been voluntarily adopted as standard terms by many cargo insurers. The most common is known as Lloyd's Marine Policy, and the cargo clauses are described as Institute Cargo Clauses ('ICC').[6]

There are three forms of ICC terms, namely:

- ICC (A);
- ICC (B); and
- ICC (C).

All ICC terms are expressly governed by English law and practice. The Marine Insurance Act 1906 (and, once it comes into force, the Insurance Act 2015) also contains detailed provisions in relation to cargo insurance policies. This chapter is written on the basis of the current law, viz the 1906 Act, although some of the changes that will be introduced by the 2015 Act are referred to at paragraph 5 below.

(a) Perils covered

ICC (A) provides coverage for loss of or damage to cargo on a comprehensive 'all risks' basis, subject to a significant number of exclusions which are enumerated in clauses 4, 5, 6 and 7. It is the most commonly used of the three ICC terms. Clause 1 of the ICC (A) terms reads:

> *[T]his insurance covers* 'all risks' *of loss of or damage to the subject-matter insured except as provided in clauses 4, 5, 6 and 7 below.* [Emphasis added]

In *British & Foreign Marine Insurance Co Ltd v Gaunt,*[7] the following observation was made on 'all risks' cover:

---

6 John Dunt, *Marine Cargo Insurance* (Informa Law, 2nd Edition, 2015) at [1.11]–[1.13].

There are, of course, limits to "all risks". They are risks and risks insured against. Accordingly, the expression does not cover inherent vice or mere wear and tear... it covers a risk, not a certainty; it is something, which happens to the subject-matter from without, not the natural behaviour of that subject-matter, being what it is, in the circumstances under which it is carried. Nor is it a loss which the assured brings about by his own act, for then he has not merely exposed the goods to the chance of injury, he has injured them himself.[8]

To come within 'all risks' cover, it is settled under English law and practice that an assured only needs to show that the loss happened as a result of a fortuity and was not caused by any of the excluded perils set out in clauses 4, 5, 6 and 7. The assured does not have to identify the specific operative peril.[9]

Coverage under the ICC (B) and (C) terms is confined to named perils, meaning that a number of specific perils are identified and cover is granted in respect of loss of or damage to cargo caused by these perils. A cargo interest who wishes to save on insurance premium by not opting for the wider ICC (A) cover would do well to ensure that the specific perils to which his cargo may be exposed are expressly covered under the ICC (B) or (C) terms.

Also available are the Institute Bulk Oil Clauses (February 1 1983), which provide cover that is specific to the carriage of bulk crude oil and liquid petroleum products. The cover provided under these clauses is better suited to oil cargoes than that provided under the ICC (A), (B) and (C) terms and is a static storage tank to static storage tank cover (rather than a warehouse to warehouse cover). It should, however, be noted that there is no cover while the cargo is in the static storage tanks at either end of the voyage; the cover only incepts once the oil leaves the tank and ends as soon as it enters the static storage tank at the other end. In addition, the Institute Bulk Oil Clauses only provide cover against the major perils plus ingress of water (much like the ICC (C) clauses). The named perils are tailored to cover risks that are more germane to oil cargoes, as follows:

1.1 loss of or contamination of the subject-matter insured reasonably attributable to
1.1.1 fire or explosion
1.1.2 vessel or craft being stranded, grounded, sunk or capsized
1.1.3 collision or contact of vessel or craft with any external object other than water
1.1.4 discharge of cargo at a port or place of distress
1.1.5 earthquake, volcanic eruption or lightning
1.2 loss of or contamination of the subject-matter insured caused by
1.2.1 general average sacrifice
1.2.2 jettison
1.2.3 leakage from connecting pipelines in loading transshipment or discharge
1.2.4 negligence of Master Officers or Crew in pumping cargo ballast or fuel
1.3 contamination of the subject-matter insured resulting from stress of weather.

7 [1921] 2 AC 41.
8 *Ibid*, at [57].
9 Howard N Bennett, *The Law of Marine Insurance* (Oxford University Press, 2nd Edition 2006) at [10.78].

Given the limited cover on offer under the Institute Bulk Oil Clauses,[10] Bulk Oil Clauses (SP-13C, January 1962) have been more popular[11] as they offer 'all risks' coverage. It should, however, be noted that cover under the latter for 'shortage and/or leakage and/or contamination' is limited to vessel casualty situations.

A draft 'all risks' Institute Bulk Oil Clauses cover provision[12] which may become popular in the future once adopted, is currently under consideration. Nevertheless, the Institute Bulk Oil Clauses remain a named perils basis insurance cover and, as a result, they are not as commonly used in the oil trading market as they might be, all risks being much preferred in the market. It is not common to insure a valuable cargo (such as oil) on any terms less favourable to the insured than the ICC (A) terms. As noted in *Geofizika v MBB*,[13] however, if the sale contract is on Incoterms without any amendments to the insurance provisions, the seller is only obliged to procure cover on the basis of the ICC (C) terms. Such terms may be considered inadequate in relation to the international sale of oil and traders should therefore be wary of the requirements of the sale contract.

Under the ICC (B) terms, coverage is limited to loss of or damage to cargo caused by:

- fire or explosion;
- the vessel or craft being stranded, grounded, sunk or capsized;
- the overturning or derailment of a land conveyance;
- collision or contact of the vessel, craft or conveyance with any external object other than water;
- the discharge of the cargo at a port of distress;
- earthquake, volcanic eruption or lightning;
- loss or damage to cargo caused by general average sacrifice;
- jettison or washing overboard and the entry of sea, lake or river water into the vessel, craft, hold, conveyance, container, liftvan or place of storage; and
- the total loss of any package lost overboard or dropped during loading onto or unloading from the vessel or craft.

In practice, however, the ICC (B) terms are rarely used.[14]

Coverage under the ICC (C) terms is largely similar to ICC (B), save that there is no coverage for:

- earthquake, volcanic eruption or lightning;
- jettison or washing overboard and the entry of sea, lake or river water into the vessel, craft, hold, conveyance, container, liftvan or place of storage; and
- the total loss of any package lost overboard or dropped during loading onto or unloading from the vessel or craft.

---

10 For further comments on the coverage offered under the Institute Bulk Oil Clauses, see N. Geoffrey Hudson and Tim Madge, *Marine Insurance Clauses* (Informa Law, 5th Edition, 2012) at Part II 4.E.

11 See Dunt (note 6 above) at [9.54]

12 *Ibid*, at Appendix 21; see also JCC Cargo Clauses website <www.lmalloyds.com/Web/market_places/marine/JCC/JCC_Clauses_Project/Cargo_Clauses.aspx> (accessed February 24 2016).

13 *Geofizika DD v MMB International Ltd and another* [2010] EWCA Civ 459.

14 Bennett (note 9 above) at [7.10].

This is the default level of insurance cover under Incoterms® and, as can be seen, cover is not comprehensive. Parties to contracts for the sale and purchase of valuable oil cargoes would therefore be well-advised to amend the Incoterms® insurance requirements such that they can be confident that terms no less favourable to the insured than the ICC (A) terms will apply during the periods when they are on risk in relation to the cargo.

As discussed at 2 above, the minimum cargo insurance cover that a seller must obtain to comply with CIF terms is ICC (C) cover or any similar clauses under English law. A buyer may request a higher level of cover under the ICC (B) or (A) terms, and/or cover for war risks.

(b) Exclusion clauses

All three sets of ICC terms expressly exclude from cover loss of or damage to or expense caused by the cargo insured occasioned by, among other things:

- wilful misconduct of the assured;
- ordinary leakage, wear and tear, or loss in weight or volume;
- unsuitability of packaging;
- inherent vice;
- a proximate cause of delay (even if the delay were caused by an insured risk); and
- insolvency or financial default of the owners, managers, charterers or operators of the vessel.

Generally, the burden of proving the application of an exclusion from cover will fall on the insurer.

Loss of or damage to or expense caused by the insured cargo arising out of war risks (ie, war, civil war, revolution, capture due to covered risks, or derelict weapons of war) or strike risks (including strikes, lockouts or persons taking part in labour disturbances, any terrorist activity, or by any person acting from a political motive) are also expressly excluded under the 'all risks' provisions of all 3 sets of ICC terms. Additionally, the ICC (B) and (C) terms have additional exclusions that are absent in ICC (A) terms.

The following discussion is germane to specific exclusions and their impact on the trade in oil cargoes.

Wilful misconduct: Both the ICC (A) terms[15] and the Marine Insurance Act 1906[16] contain exclusions of cover where the loss is attributable to the wilful misconduct of the insured. The general purpose of this is to exclude from cover events which the insured brings about himself (whether or not by a criminal act) in order to create a claim under the insurance policy.

The most commonly cited example in this regard is the deliberate scuttling of a ship by its owner in order to make a claim under the insurance policy.[17] In the

15 See for example, Clause 4.1.
16 Marine Insurance Act 1906, section 55(2)(a).
17 See for example, *Brownsville Holdings Ltd v Adamjee Ins Co Ltd (The Milasan)* [2000] 2 Lloyd's Rep 458.

context of a cargo insurance policy covering an oil cargo on board a vessel, it is relatively unlikely that this exclusion will arise (unless the cargo owner is also the shipowner), as the cargo owner generally passes the cargo into the hands of a carrier during the period when the insurance applies. In this respect, even if the carrier deliberately casts away the vessel during the voyage, the wilful misconduct exclusion will generally not apply, unless the cargo owner is complicit in that action.[18]

Ordinary leakage, wear and tear, loss in weight or volume: It is difficult to foresee a situation in which damage to an oil cargo might be described as 'wear and tear', as situations where the cargo is unstable, such that it is incapable of withstanding the contractual voyage and remaining within the contractual specification,[19] are more likely to be viewed as inherent vice of the cargo.

Leakage that can be expected to occur in the ordinary course of the carriage is excluded from cover; though again it is unlikely that any leakage of an oil cargo can be expected during the sea carriage, as this would most likely involve a serious breach of the shipowner's obligations to avoid marine pollution.

Likewise, ordinary evaporation loss that can be expected during a voyage (see also section 4.2 below for discussion) is not covered. Many marine cargo insurers will, however, provide an extension of cover in the cargo insurance policy in respect of such potential losses. In practice, any extensions of cover granted for this type of loss will generally be subject to a deductible that is equivalent to the evaporation loss to be expected in relation to the particular cargo carried.

Inherent vice: See section 4.2 below for discussion.

Delay: This exclusion expressly applies to delay, whether or not the delay originally arises from an insured peril;[20] the application of the exclusion is, however, mitigated by, firstly, the fact that general average charges remain payable under ICC (A), clause 2 regardless of delay, and, secondly, the ability of the insured to exercise an option under ICC (A), clause 9 to seek an extension of the insurance cover where the contractual voyage is terminated before the vessel reaches the discharge port.

Nevertheless, losses directly arising from delay, such as a fall in the market value of an oil cargo, will not be covered.

Insolvency or financial default of the owners etc of the vessel: This exclusion applies where the insured is (or should be) aware, at the time when the vessel is loaded,[21] that insolvency or financial default of the vessel's owners, managers, charterers or operators might prevent the normal prosecution of the voyage. As a

18 See also *Shell International Petroleum Co Ltd v Gibbs (The Salem)* [1983] 1 Lloyd's Rep 342, in which Lord Roskill held that the oil cargo remaining on board the *Salem* at the time she was scuttled was lost by perils of the sea, in circumstances where it was far from certain that the fraud culminating in the scuttling would succeed at the time when the voyage was commenced.

19 As in, for example *KG Bominflot Bunkergesellschaft Für Mineralöle Mbh & Co v Petroplus Marketing Ag (The 'Mercini Lady') (No 2)* [2012] EWHC 3009 (Comm).

20 See ICC (A), clause 4.5

21 See FD Rose, *Marine Insurance: Law and Practice* (Informa Law, 2nd Edition, 2012) at [18.58].

result, in order to ensure that he does not fall foul of this exclusion, an insured should carry out some due diligence before selecting a carrier for his cargo. In order to prevent potential unfairness to third-party purchasers of cargoes, who receive an assignment of insurance originally procured by a cargo owner higher up in the chain of sale contracts, this exclusion does not apply to such purchasers where they have bought the cargo in good faith.

Unseaworthiness: Although unseaworthiness is not an exclusion under the ICC terms, there is an implied warranty of seaworthiness under section 39 of the Marine Insurance Act 1906 for voyage policies (cargo insurance being provided on a voyage basis), which, if breached, could provide underwriters with a defence to a claim under a cargo insurance policy. That implied warranty is waived under the ICC (A), (B) and (C) terms, clause 5.2, save where the unseaworthiness occurs with the privity of the assured or its servants. It follows that unseaworthiness will not generally give rise to a reason to deny cover in a policy concluded on ICC terms.

(c) Duration of cover

The transit clauses in all three sets of ICC terms (formerly referred to as 'warehouse to warehouse cover') extend coverage for the sea voyage to include risks while the cargo is transported from the inland warehouse or place of storage to the time and place of loading (ie, the place named for the commencement of the transit), and the risks from the time and place of cargo discharge to the delivery of the cargo at the final warehouse or place of storage (ie, the named destination).[22]

This is subject to an overriding time limit of 60 days after the completion of discharge from the vessel at the final port of discharge. The cover ceases on the expiration of that 60-day time limit, even though the goods may not have reached the named destination warehouse. If, prior to the expiration of the 60-day time limit, the cargo is forwarded after discharge from the vessel to a destination other than that named in the policy, coverage terminates on the commencement of the inland transit.[23]

The 60-day coverage is extremely helpful if the assured is (for whatever reason) unable to transport the cargo to the designated storage tank or vessel. It could then take advantage of the 60-day time limit to arrange alternative coverage for the cargo before coverage under the existing policy terminates.

3.2 Protection and indemnity ('P&I') cover

As a general rule of thumb, cargo insurance provides cover for first-party losses, while P&I insurance (generally purchased by shipowner members who enter their vessels with a mutual P&I Club) is designed to indemnify a shipowner or charterer in respect of the discharge of legal liabilities incurred in the operation of the vessel.[24]

---

22 See, for example, Clause 8.1 of the ICC (A) terms.
23 See clause 8.2 of the ICC (A) terms.
24 Steven J. Hazelwood, David Semark, *P.&.I Clubs Law and Practice* (Taylor & Francis, 4th Edition, 2013) at [1.1].

The following risks are covered by P&I insurance:[25]

- collision with other vessels;[26]
- property not on board an entered vessel[27] or on board an entered vessel other than cargo;[28]
- liability to persons other than seamen;[29]
- illness, injury and death of seamen;[30]
- diversion;[31] and
- towage.[32]

Most relevant with regard to oil cargo, P&I insurance covers cargo risk[33] and pollution risk.[34] The former will be significant for oil cargo owners who do not own or charter the vessel, and will therefore be seeking an indemnity from the shipowners or charterers for any loss of or damage to the cargo ensuing from the carriage. The cover for pollution risks will be significant for oil traders who own both the oil cargo and the vessel (or who bareboat charter the vessel), and who may be liable for loss or expense arising out of pollution. The general position is, however, that the polluter pays and that it is the ship which is responsible for clean-up costs. It follows that traders who only own the cargo are not usually pursued for oil pollution liabilities, although that can happen in some jurisdictions, such as certain US states.

(a) Cargo claims

Subject to a number of exclusions,[35] P&I insurance will cover liabilities, costs and expenses resulting from claims made by cargo interests for loss of or damage to cargo, to the extent of the cargo having been carried by the entered vessel. It was decided in a London arbitration that P&I cover in respect of cargo liabilities did not cover a member's liabilities to a charterer in respect of salvage services rendered to bunkers owned by the charterer.[36] Cover is triggered where the member has not properly performed the carriage – there will, however, be no cover where the member has failed to perform the carriage at all.[37]

The cargo interests will usually be covered by cargo insurance (see the previous discussion on the ICC terms), and the first point of indemnity will usually be from the cargo insurers to the cargo interests. Provided that cargo insurers can prove legal

---

25 See, for example, rule 9 of the London P&I Club Rules 2015–2016, effective February 20 2015.
26 *Ibid*, rule 9.13.
27 *Ibid*, rule 9.14.
28 *Ibid*, rule 9.22.
29 *Ibid*, rule 9.2.
30 *Ibid*, rules 9.3 and 9.4.
31 *Ibid*, rule 9.7.
32 *Ibid*, rule 9.16.
33 *Ibid*, rule 9.19.
34 *Ibid*, rule 9.15.
35 *Ibid*, rule 9.19.2.
36 London arbitration No 17/86 (Summary-Lloyd's Maritime Law Newsletter (1986) 181 LMLN 3), accessible at Lloyd's Maritime Law Newsletter website www.lmln.com/admiralty/salvage/article19252.ece (accessed 24 February 2016).
37 Hazlewood & Semark (note 24 above) at [10.55]. See also, for example, rule 9.19.1 of London P&I Club Rules 2015–2016, which states that that cover applies to cargo which is "intended to be or [is] being or having been carried in an entered Ship".

liability on the part of the shipowner in relation to the loss of or damage to cargo, cargo insurers may then bring subrogated recovery proceedings against the shipowner. Alternatively, should the cargo interest be uninsured or self-insured or the cargo insurance in place cannot be called upon, a claim may be advanced directly against the shipowner. In either case, P&I insurance will *prima facie* respond to indemnify the shipowner. It should, however, be noted that most P&I clubs include a 'pay to be paid' rule in their terms of cover with the shipowner, the effect of which is generally to exclude direct rights of action against the P&I Club in circumstances where the shipowner is unable to pay the claim itself.[38]

Subject to the specific rules of the P&I club that the vessel is entered with, cover is generally provided for the following:

- Provided that there was a breach by the member or his agent/servant of his obligation properly to load, handle, stow, carry, keep, care for, discharge or deliver the cargo, or a breach arising out of the unseaworthiness or uncargoworthiness of the entered vessel, the rule covers claims for short delivery, unseaworthiness, improper stowage, inadequate dunnaging, inadequate heating or ventilation;[39]
- The additional costs and expenses incurred by the member in discharging or disposing of damaged or worthless cargo, if (or to the extent that) the member has no right to recover these costs from any other party;[40]
- Any liability under a bill of lading (or other form of contract approved by the managers and providing for the carriage) partly performed by the insured vessel and partly by other means of transport, including handling outside the dock areas of the ports of loading and discharge;[41]
- Liabilities, costs and expenses of shipowners in respect of loss of or damage to or responsibility in respect of cargo or other property for which the shipowner may be liable under indemnities or undertakings to operators of cranes, lighters, craft or other appliances used in connection with the loading or discharging of cargo, provided that such indemnities or undertakings have received the prior approval of the club;[42] and
- Liabilities, costs and expenses of shipowners in respect of indemnities to stevedores or others engaged in connection with the storage, custody, handling or transport of the cargo.[43]

In the event that a cargo owner does not also own the vessel in which the cargo is carried, the cargo owner may not have a direct right of claim or recovery against

---

38 See *Firma C-Trade SA v Newcastle Protection and Indemnity Association (The 'Fanti'), Socony Mobil Oil Co Inc and others v West of England Ship Owners Mutual Insurance Association Ltd (The 'Padre Island ') (No 2)* [1989] 1 Lloyd's Rep 239.

39 Hazlewood & Semark (note 24 above) at [10.55]. See also, for example, rule 9.19.1.1 of London P&I Club Rules 2015–2016.

40 Hazlewood & Semark (note 24 above) at [10.56]. See also, for example, rules 9.19.1.2 and 9.19.1.3 of London P&I Club Rules 2015–2016.

41 Hazlewood & Semark (note 24 above), at [10.57]. See also, for example, rule 9.19.1.4 of the London P&I Club Rules 2015–2016.

42 Hazlewood & Semark (note 24 above) at [10.58].

43 *Ibid.*

the P&I Club with which the vessel is entered. Recourse may be made in this case via the vessel owner who is a member of the P&I Club, but will be subject to the 'pay to be paid rule' referred to previously, such that recovery actions against an insolvent vessel owner are unlikely to result in payment of the claim and will not necessarily trigger an obligation for the P&I Club to pay.

(b) Pollution

Shipowners face significant exposure to liability claims in respect of pollution caused by oil cargoes carried on board their vessels. International conventions such as the International Convention on Civil Liability for Oil Pollution Damage ('CLC'), domestic legislation and the common law seek to regulate and, in some cases, cap the liabilities, but other liability regimes have been voluntarily assumed by shipowners in accordance with schemes such as the Small Tanker Oil Pollution Indemnification Agreement ('STOPIA') and the Tanker Oil Pollution Indemnification Agreement ('TOPIA'). The liabilities from oil pollution encompassed in these conventions and voluntary agreements are insured by P&I clubs, subject generally to a limit of liability of $1 billion per entered vessel per accident or occurrence.[44]

P&I clubs will indemnify their members in respect of legal damages and compensation payable in respect of pollution incidents in the members' capacity as shipowners or charterers of the entered vessel. Any liabilities incurred in any other capacity (eg, as cargo owners) will not be covered. The latter distinction will be relevant where the members own both the vessel and the cargo (eg, some oil majors).

Under English law, an owner's liability to compensate for oil pollution is encompassed by the International Convention on the Establishment of an International Fund for Compensation for Oil Pollution Damage, the CLC and the common law.

It is beyond the scope of this chapter to examine the international conventions and common law positions in detail;[45] what follows is a general overview. Where P&I insurance is concerned, the CLC requires compensation to be paid to victims of oil pollution by shipowners as the carriers of oil and mandates, in this regard, that shipowners should take up compulsory liability insurance to ensure that they have sufficient funds to meet claims under the conventions. This requirement can be met by entry into a P&I Club.[46]

To illustrate the ambit of P&I cover, let us assume hypothetically that there has been a casualty involving oil pollution. In such a situation, P&I cover will commonly respond in the following manner.

- Following the casualty, the local governmental authorities of the coastal state threatened or affected by any oil pollution will usually undertake work to prevent or minimise the damage and carry out clean-up operations. Such expenses would likely qualify for reimbursement from the P&I Club, along with any costs and expenses that the shipowner may himself incur for the

44 Rule 11.3 of the London P&I Club Rules 2015–2016.

45 A comprehensive review of the law in this area can be found in Colin de la Rue & Charles B Anderson, *Shipping and the Environment* (Informa Law, 2nd Edition, 2009).

46 Hazlewood & Semark (note 24 above), at [10.225].

same purpose. Unless agreed otherwise by the club, such expenses will not be recoverable if they would have been recoverable as general average expenses had the contract of carriage of oil been subject to the York-Antwerp Rules 1994.[47]

- Reimbursement of expenses incurred to comply with orders given by the authorities of the state affected by the pollution incident. Not all such expenses will fall within P&I Club cover, eg, where compliance with the order is considered part of the normal operation or maintenance of the vessel. Some expenses could also be categorised as salvage charges, or as expenses under a Pollution Hazard Clause, or as general average expenses in a marine hull policy.
- A shipowner may enter into various types of contracts involving the provision of an indemnity to the other party in respect of oil pollution. An example would be a situation in which a stricken vessel wishes to enter a port and the port (which is concerned by the threat of pollution) makes the granting of such permission contingent on an indemnity being provided by the owner for any pollution caused by the vessel while inside the port. P&I cover would usually expressly provide cover in such situations, notwithstanding a general cautiousness about covering members, for contractual liabilities voluntarily entered into outside of ordinary contracts of carriage of goods or customary towage.
- Indemnification of criminal fines payable by shipowners and crew. An example would be the English Merchant Shipping Act 1995, which (read in conjunction with the regulations promulgated pursuant to European Union Directive 2005/35/EC on Ship Source Pollution[48]) provides, first, for a shipowner or master to be prosecuted in respect of any discharge or escape of oil;[49] secondly, for fines to be payable for any failure to report a discharge or escape of oil;[50] thirdly, to keep on-board records of discharges and escapes of oil;[51] fourthly, to obey directions of the Secretary of State following an accident;[52] and, finally, to produce ship's records.[53]

3.3 Kidnap and ransom ('K&R') cover

K&R cover was developed in response to the increasing costs of ransoms paid for kidnapped crew members and the general unwillingness of P&I clubs to participate in the payment of these claims.

The insurance is intended principally to provide cover for:

- the ransom payment;
- the loss of the ransom amount while it is being taken to the place of payment; and

---

47 Hazlewood & Semark (note 24 above), at [10.238].
48 Effect is given to the directive by way of the Merchant Shipping (Implementation of Ship-Source Pollution Directive) Regulations 2009 (SI 2009 No 1210).
49 Section 131 of the Merchant Shipping Act 1995 (Cap 21).
50 *Ibid*, section 135.
51 *Ibid*, section 142.
52 *Ibid*, sections 137 and 139.
53 *Ibid*, sections 259 and 260.

- fees and expenses peculiar to the nature of this cover (such as the costs of security experts).

It also provides cover for:

- ancillary expenses that may be incurred in negotiations for the release of the crew, vessel and cargo, such as expenses in relation to the appointment of public relations consultants, specialist response consultants, interpreters and independent negotiators; and
- liability arising out of or in connection with a piracy event.

An example of payments covered would be legal expenses incurred and liability payments (whether arising from judgments or settlements) paid out pursuant to claims or actions brought by crew members for damages suffered as a result of the insured kidnap event. It is generally a fundamental breach of such policies to disclose publicly that the cover exists, for the obvious reason that this tends to inflate ransom demands.

4. Legal issues in oil cargo insurance

4.1 Contamination: issues in loading oil cargo onto and discharging it from tankers

A tanker is a vessel designed for the carriage of liquid in bulk, with cargo spaces consisting of several tanks. The products carried are varied and include crude oil, refined products and liquid gas. The size and capacity of tankers range from small coastal vessels to very large crude carriers, sometimes in excess of 500,000 tonnes deadweight.

Tankers load their cargo by gravity from the shore or by shore pumps and discharge it using their own pumps. Contamination can result in serious damage during the loading and discharging processes and brings into focus issues relating to the seaworthiness and cargoworthiness of the vessel.

Contamination resulting in an oil cargo taking on a more dangerous nature than originally expected was considered in *Mediterranean Freight Services Ltd v BP Oil International Ltd (The Fiona).*[54] In that case, there was an explosion in the No. 1 tank of the plaintiff's tanker while the defendant charterer was preparing to discharge her cargo of fuel oil. Immediately before the explosion, two surveyors had approached the No. 1 tank to take ullage soundings and temperature readings. One of the surveyors inserted an unearthed electronic temperature probe and the tank exploded, resulting in the death of the other surveyor and significant damage to the tanker.

The bill of lading was subject to the Hague-Visby Rules ('the Rules') which were incorporated by virtue of the Carriage of Goods by Sea Act 1971. The plaintiff issued proceedings seeking damages as well as an indemnity for any cargo or personal injury claim on the basis that the fuel oil cargo was of a dangerous character, so that the plaintiff was entitled to relief under article IV.6 of the Rules. The defendant

54 [1993] 1 Lloyd's Rep 257 (first instance) and [1994] 2 Lloyd's Rep 506 (Court of Appeal).

denied liability and claimed that the plaintiff breached article III.1 of the Rules by reason of its failure to discharge from the vessel remnants of the previous cargo of volatile petroleum condensate. The contamination rendered the tanker unseaworthy. The question before the court was whether the plaintiff had failed to exercise due diligence to maintain the seaworthiness of the tanker.

The first instance court held[55] that the blend of the fuel oil cargo had a tendency to give off light hydrocarbon gases in tank head spaces which sometimes approached (and occasionally exceeded) the lower explosive limit irrespective of flashpoint temperature, and was therefore dangerous. This was one of the causes of the explosion. The plaintiff was, therefore, in principle entitled to rely on article IV.6 of the Rules, as this constituted dangerous cargo. The court also found that the oil in No. 1 tank was contaminated by the highly volatile condensate left on board from the previous voyage, and that this was a concurrent and dominant cause of the explosion. The plaintiff had therefore breached article III.1 by failing to exercise due diligence to render the vessel seaworthy. This breach provided the defendant with a defence to the plaintiff's claim in respect of dangerous goods, pursuant to article IV.6 of the Rules.

The Court of Appeal agreed with the first instance court and dismissed the appeal.[56] It held that the case was governed by the general rule laid down in the line of cases including *Canada Steamship Line Ltd v The King*[57] which holds that an indemnity will not lie in respect of loss due to a person's own negligence unless adequate and clear words were used or unless the indemnity could have no reasonable meaning or application unless so applied.

Proving where contamination arises, however, can be problematic. In *Fal Oil Co Ltd v Petronas Trading Corporation Sdn Bhd*,[58] a claim was advanced by the plaintiff for demurrage arising from delay in the discharge of a cargo of oil, but contamination issues were central to the consideration of the claim. Under the sale contract, the oil had a tolerance level of up to one per cent of water. Prior to loading, the oil was tested and the water and sediment content were found to be within the tolerance range. Subsequent to a ship-to-ship transfer ('STS') of the oil to the vessel, *The Devon*, a composite sample was again taken from each of the ship's four tanks and was found to be within range, although the surveyor noted traces of water in the tanks. Discharge in Malaysia was halted when it was found that samples taken had a water content of up to 31.4%, comprising mostly sea water.

The Commercial Court held in favour of the plaintiff and was of the view that the central issue was whether the cargo transferred to *The Devon* at the point of STS contained the water contamination. The burden of proof was on the defendant to show (on a balance of probabilities) that the contamination occurred prior to the STS transfer. There was conflicting expert evidence in this case, and the court was of the view that there had probably been a deliberate switch of sea water for cargo oil. The difficulty was in concluding whether the switch occurred before or after the STS

55 [1993] 1 Lloyd's Rep 257 and LMLN 345 (23 January 1993).
56 [1994] 2 Lloyd's Rep 506.
57 [1952] AC 192.
58 [2003] EWHC 2225 (Comm).

transfer. Given that the defendant bore the burden of proof, it followed that it had not made out its case.

4.2 Shortage

Prior to 1973, oil cargo shortage claims were largely unheard of outside of major casualties in which large amounts of oil cargo had been lost. Following the rise in crude oil prices post-1973, oil companies and cargo underwriters became more concerned with oil cargo losses.[59]

Claims for short delivery of oil cargoes have since given rise to complex disputes. The consensus is that the principles of law are clear but, as always, it is the application of facts and the evidence to the principles that gives rise to difficulties. Each case, therefore, turns on its own facts.[60] In *Amoco v Parpada Shipping (The George S)*,[61] the cargo owners made a claim for short delivery in respect of a cargo of crude oil. The first instance judge held that the discrepancy could well have arisen as a result of a theoretical possibility that oil remained in the pipeline (which was over a mile long) between the ship and the shore terminal (in circumstances where delivery took place at the ship's permanent manifold connected to the pipeline and not at the shore terminal), and that the cargo owners had therefore failed to prove that there was a shortage. The claim was accordingly dismissed.

The decision was reversed on appeal, on the basis that the first instance judge had not given adequate weight to, firstly, the fact that the vessel figures at the load and discharge ports (when compared on a 'like with like' basis) and the shore figures at the load and discharge ports (also compared 'like with like'), both pointed to a substantial short delivery, and secondly, the shipowner's explanation of the apparent shortage, which was considered inadequate.

A summary of the legal framework for a short-delivery claim in relation to oil cargoes is as follows.[62]

- The claimant has the burden of proving, on a balance of probabilities,[63] that the vessel delivered less than the cargo originally loaded. The standard of proof applies even if the short-delivery claim involves an accusation of theft by the owner or the crew.[64]
- The bill of lading is relied upon by third parties in relation to the facts stated in it and as a result, even where it is only *prima facie* evidence of the cargo quantity, evidence of the cargo quantity in the bill of lading will not be easily displaced. To do so will require convincing proof. Moreover, evidence in the

---

59 Intertanko, *Oil Cargo Losses and Problems with Measurement* (Intertanko, 2nd Ed, 1996), at p 5.

60 Julian Cooke, Andrew Taylor, John D Kimball, David Martowski & LeRoy Lambert, *Voyage Charters* (Informa Law, 4th Edition, 2014) at [60.5].

61 [1989] 1 Lloyd's Rep 369 ("*The George S*").

62 Cooke, Taylor, Kimball, Martowski & Lambert, (note 60 above) at [60.5].

63 Lord Donaldson MR in *The George S* (note 61 above) suggested that after a *prima facie* case of shortage has been established, a shipowner must prove beyond reasonable doubt that he delivered all that he received in order to escape liability. However, in *Anonima Petroli Italiana v Marlucidez Armadora SA* (The Filiatra Legacy) [1991] 2 Lloyd's Rep 337 ("*The Filiatra Legacy*"), the Court of Appeal clarified that Lord Donaldson MR's remarks were confined only to a situation where the onshore evidence points inexorably to a short delivery.

64 See *The Filiatra Legacy* (*Ibid*). The likelihood that the owner or crew would collaborate on a theft (in the scale and manner involved) will, however, be one of the factors to be weighed.

bill of lading may give rise to estoppel arguments against the vessel where the bill of lading holder relies to his detriment on statements contained in the bill.

- The responsibility of the shipowner begins at loading and ends on the discharge of the cargo. Any intermediate measurements of the quantities of oil are "merely means of determining quantities at those crucial points."[65] Therefore, unless a meter were installed in the loading or discharge pipe, there would be no measurement of quantity at the crucial point. The loading and discharged quantity would then have to be inferred from the vessel's tank ullages and the shore figures.[66]
- If the oil cargo owner establishes short delivery, the burden shifts to the shipowner to show the short delivery was caused by an event for which he is not contractually responsible. The standard of proof in these circumstances appears to be proof "beyond all reasonable doubt"[67] that a short delivery was impossible. The author must, however, respectfully query why the criminal standard of proof should apply here, when only the civil standard (proof on a balance of probabilities) generally applies in civil cases.
- It is common practice for the owner not to be liable for shortage within a certain range (usually between 0.3% and 0.5%, to allow for evaporation). The charter may also contain a clause permitting the charterer to withhold from freight the value of short-delivered cargo – the permissible shortage should then be taken into account in calculating the amount short-delivered.[68]

The practicalities of bulk oil shortage claims have naturally caused issues for insurers, in particular 'paper losses' caused by:

- differences in the sampling methods for water content at load and discharge ports;
- water settling out during the voyage;
- differing quantity measuring techniques used at load and discharge ports; or
- shore lines being part empty at commencement of discharge, but full on its completion.

The Institute Bulk Oil Clauses include the following provision in relation to recoverable shortage and leakage claims:

15. Claims for leakage and shortage recoverable under this insurance are to be adjusted as follows:-

15.1 The amount recoverable shall be the proportionate insured value of the volume of oil lost, to be ascertained by a comparison of the gross volume certified as having left tanks for loading on to the vessel with the gross volume certified as having been delivered to tanks at the termination of the transit, except that where the contract of sale

65 Per Lord Donaldson MR in *The George S* (note 61 above) at p 371.
66 Cooke, Taylor, Kimball, Martowski & Lambert, (note 60 above) at [60.6].
67 Per Lord Donaldson MR in *The George S* (note 61 above) at p 376.
68 See *Trafigura Beheer BV v Navigazione Montanari SpA (Valle di Cordoba)* [2015] EWCA Civ 91 in relation to the operation of such clauses.

is based on weight and not on volume the amount recoverable may be calculated on a weight basis from such certified quantities.

The term "gross volume" in this Clause 15.1 means total volume without deduction of sediment and water content and free water, except to the extent that the amount of water can be shown by the Assured to have increased abnormally during the insured transit as a result of the operation of a risk covered by this insurance.

15.2 Adjustment shall be made to the calculation under 15.1 above to eliminate any change in volume caused by variation in temperature and any apparent change in quantity arising from the use of inconsistent procedures in determining the certified quantities.

15.3 Where this insurance provides for an excess to be applied to claims for leakage or shortage, such excess shall be deemed to include ordinary loss in weight or volume except when caused by variation in temperature or settling out of water. Where there is no such provision, the amount recoverable in accordance with Clauses 15.1 and 15.2 shall be subject to reduction for any ordinary loss excluded by Clause 4.2 above.

This wording is intended to exclude cargo losses that are bound to occur during the voyage; clause 15.2 in particular is directed at the issue of evaporation loss and 'paper losses' resulting from differing cargo measurement techniques.

4.3 Inherent vice

While an oil cargo is in the custody of the carrier, it is expected that there will be certain inevitable in-transit loss due to the natural characteristics of the cargo. Such loss, due to the nature or chemical characteristics of the oil cargo, opens the door to arguments that the loss is due to the inherent vice of the oil cargo and therefore not recoverable under the cargo policy.

The ICC terms provide that in no case shall the insurance cover "loss damage or expense caused by inherent vice or nature of the subject-matter insured".[69]

'Inherent vice' has been defined as a loss "which is proximately caused by the natural behaviour of the subject-matter insured, being what it is, in the circumstances in which it was expected to be carried".[70] It is further explained that 'inherent vice' is not descriptive of the loss itself, but of the risk of deterioration of the goods shipped as a result of their natural behaviour in the ordinary course of the contemplated voyage, without the intervention of any fortuitous external accident or casualty.[71]

Albeit outside the context of an oil cargo, the leading modern case on inherent vice is *Global Process Systems Inc and Another v Syarikat Takaful Malaysia Berhad (The Cendor Mopu)*,[72] which concerned the scope of the inherent vice exclusion in a marine insurance policy.

The *Cendor Mopu* was an oil rig which was laid up in Texas and purchased by the assured for conversion into a mobile offshore production unit for use in Malaysia. The wet tow of the rig (with its legs jacked up) by a barge from Texas to Malaysia was insured on ICC (A) 'all risks' terms. *En route*, the repeated bending of the legs of the

---

69 ICC (A), (B) and (C) terms, clause 4.4.
70 *Soya GmbH v White* [1982] 1 Lloyd's Rep 136 at 149 (Court of Appeal).
71 *Soya GmbH v White* [1983] 1 Lloyd's Rep 122 at 126 (House of Lords).
72 [2009] Lloyd's Rep Plus 47.

rig by the motion of the barge during the tow caused fatigue cracking, and the eventual breakage and loss of three of the legs.

Insurers rejected the claim on the basis that the loss was inevitable (ie, not a 'fortuity'), or was proximately caused by inherent vice. The first instance court found that the loss was "very probable" but not inevitable because a "leg breaking wave" was required to generate the final fracture. The defence of inherent vice succeeded, however, on the ground that the legs of the rig were not capable of withstanding the normal incidents of the voyage (including the weather reasonably to be expected).

The first instance decision was reversed on appeal to the Court of Appeal.[73] The Court of Appeal held that the proximate cause of the loss was an insured peril in the form of the leg breaking wave, and not inherent vice.

On appeal by the insurers to the Supreme Court,[74] it was held that the proximate cause was not inherent vice but perils of the sea. The court opined that loss was not inevitable in the ordinary course of the voyage: the breakage only occurred under the influence of a wave of a direction and strength catching the first leg just at the right moment, thereby leading to increased stress on, and the ensuing collapse of, the other two legs.

This imports a high standard in relation to inherent vice. It stands to reason that the operation of any fortuitous external accident or casualty will suffice to prevent the loss being attributed to inherent vice; the fact that the rig's legs were not capable of withstanding the normal incidents of the voyage and anticipated weather conditions did not make inherent vice the proximate cause.

4.4 Piracy

On November 15 2008, the VLCC *MV Sirius Star*, carrying two million barrels of crude oil, was hijacked by Somali pirates. At the time of the attack, the vessel was carrying 25 crewmen. It was estimated to be worth approximately $150 million and her cargo worth at least $100 million. The vessel was released on January 9 2009, after the payment of a $3 million ransom. Since then, other VLCCs that have been hijacked and ransomed by pirates at the time of writing are the *Samho Dream*, the *Maran Centaurus* and the *Irene SL*.

The costs directly related to piracy have traditionally been covered by a cargo insurance policy for the goods being shipped and a hull and machinery policy for the vessel itself. Shipowners whose vessels ply volatile areas additionally purchase K&R insurance (see section 3.3 above). Without this, injury or death to a crew member would likely be covered by P&I insurance.

Through the years, piracy has fluctuated between being treated as a marine risk and a war risk. It is now included in standard clauses for hull and cargo clauses, with an express provision that piracy is not excluded by the war exclusion clause.

Under the ICC (A) terms, piracy is included in the 'all risks' cover, but is excepted in the war exclusion clause. The ICC (B) and (C) terms do not list piracy as an insured peril, and neither is it excepted from the war exclusion clause.

73 [2010] Lloyd's Rep IR 221.
74 [2011] UKSC 5.

The ICC (A) terms provide:

6. In no case shall this insurance cover loss damage or expense caused by

6.1 capture seizure arrest restraint or detainment (piracy excepted), *and the consequences thereof or any attempt thereat* [Emphasis added]

The relevant exclusion in the ICC (B) and (C) terms provides:

6. In no case shall this insurance cover loss damage or expense caused by

...

6.2 capture seizure arrest restraint or detainment, and the consequences thereof or any attempt thereat.

Therefore, given that the war exclusion in the ICC (A) terms does not exclude piracy, this risk should be covered as a marine peril under the 'all risks' cover. For the ICC (B) and (C) terms, piracy is excluded under the war exclusion. The assured may wish to purchase additional cover under the Institute War Clause with ICC (B) and (C) cover. Even so, there is arguably still no cover for piracy. The Institute War Clauses (Cargo) provides as follows:

RISKS COVERED

1. This insurance covers, except as provided in Clauses 3 and 4 below, loss of or damage to the subject-matter insured caused by

1.1 war civil war revolution rebellion insurrection, or civil strife arising therefrom, or any hostile act by or against a belligerent power

1.2 capture seizure arrest restraint or detainment, arising from risks covered under 1.1 above, *and the consequences thereof or any attempt thereat*

1.3 derelict mines torpedoes bombs or other derelict weapons of war.

[Emphasis added]

As can be seen, cover for capture, seizure and detainment in Clause 1.2 is subject to the perils arising from Clause 1.1, which do not include piracy.

The Institute Bulk Oil Clauses are on a named-perils basis, but the war exclusion clause does not apply to piracy:

6. In no case shall this insurance cover loss damage or expense caused by

...

6.2 war, civil war, revolution, rebellion, insurrection or civil strife arising therefrom, or any hostile act by or against a belligerent power

6.3 capture, seizure, arrest, restraint or detainment (piracy excepted), *and the consequences thereof or any attempt thereat*

6.4 derelict mines, torpedoes, bombs or other derelict weapons of war.

[Emphasis added]

Given, however, that piracy is not a peril named in the Institute Bulk Oil Clauses, its exception in the war exclusion clause will not promote positive coverage (the Institute Bulk Oil Clauses being on a named peril basis, rather than an 'all risks' basis). Nevertheless, it does leave open the option of claiming ransom payments as General Average expenses, since cover for General Average is arguably not affected by piracy not being a named peril, so long as it is not an excluded peril.

4.5 Pollution

Spills from vessels are one of the major sources of water pollution from shipping.

They can range from oil spills from oil tankers (which account for the most serious incidents) to accidental discharges of oil and fuel from smaller craft.

Under the CLC regime,[75] a shipowner is liable for pollution damage caused by the escape or discharge of persistent oil from his vessel, even if the pollution was not due to his actual fault. There is a possibility of limiting his liability by reference to the tonnage of the particular ship involved.[76]

A victim will first look to the registered shipowner and his CLC insurer (usually a P&I Club that the vessel is entered with) for the initial tier of compensation. Claims can be brought directly against the CLC insurer.

The CLC regime applies to any seagoing vessel or craft of any type whatsoever, constructed or adapted for carrying 'persistent' hydrocarbon mineral oil in bulk as cargo. The term 'persistent' oils is generally taken to include crude oil, fuel oil, heavy diesel oil and lubricating oil: these comprise oils which (due to their chemical composition) are slow to dissipate naturally, are likely to spread and require cleaning up. Non-persistent oils (eg, gasoline, light diesel oil and kerosene) evaporate quickly and do not require cleaning up.

Claims may arise for the following items:

- direct damage to property, or consequential loss or expense incurred in cleaning up the pollution or contamination of property;
- economic loss suffered through prevention of normal business operations;
- costs and expenses incurred to reinstate recovery of the environment; and
- legal and other professional costs incurred by the assured in claim presentation.

For a claim for pure economic loss to be allowed, there needs to be a sufficient nexus between the oil contamination and the loss or damage sustained. The following is taken into account in determining whether a nexus exists:

- geographic proximity between the claimant's activity and the contamination;
- the degree of economic dependence on the affected resource;
- the extent to which a claimant had alternative sources of supply or business opportunities; and
- the extent to which a claimant's business formed an integral part of the economic activity within the area affected by the spill.

The burden of proof is on the shipowner to show that he is not liable for the oil spill/pollution. This can be achieved if the registered shipowner can show that the damage or the threat of contamination arose from the following events provided for by article III:

- an act of war, hostilities, civil war, insurrection or a natural phenomenon of an exceptional, inevitable and irresistible character; or

---

75 1992 version.

76 The 1992 Fund Convention is supplementary to the 1992 CLC and establishes a regime for compensating victims when compensation under the CLC is not available or is inadequate. Discussion of this area is not within the scope of this chapter.

- was wholly caused by an act or omission done with intent to cause damage by a third party (eg, sabotage/terrorism); or
- was wholly caused by the negligence or other wrongful act of any government or other authority responsible for the maintenance of lights or other navigational aids in the exercise of that function.

An insurer is absolved from liability if it can show that:

- any defences affecting the shipowner's liability are applicable; or
- the pollution damage was due to the wilful misconduct of the shipowner himself.

5. The new Insurance Act

The UK Parliament passed the Insurance Act 2015 on February 12 2015. This legislation introduces the most significant changes to English commercial insurance law in at least 100 years. The amendments are expected to come into force in August 2016. With this in mind, some leading insurers have started acting on the basis that the new laws are already effective.

The new law will apply to all insurance, including oil cargo insurance, with the exception of consumer insurance. It will also apply to reinsurance. There are also a considerable number of amended provisions. A summary of the key changes is as follows.

- With respect to the duty of fair presentation, there is now a greater emphasis on the insurer to ask questions about the risk and to make clear what information it requires. The insured must disclose all material circumstances about the risk or give the insurer sufficient information to put it on notice that it needs to make further enquires for the purposes of revealing all the material circumstances about the risk. The insured is obliged to disclose information in a manner that is reasonably clear and accessible to a prudent underwriter (ie, no data dumping) and not to misrepresent material information.
- The single remedy of avoidance from inception for a breach by the insured of its duty to disclose the risk at placement (ie, void *ab initio*) is abolished. It will be replaced by a new system of graduated remedies, based on what the insurer would have done had a fair presentation been made. For example, if the underwriter would not have written the risk at all, avoidance would be an appropriate remedy. If the underwriter would have written the risk but on different terms, these differing terms of the contract may be imposed. If the breach of duty was deliberate or reckless, the insurer may avoid the contract and keep the premium regardless.
- A breach of a warranty will no longer automatically discharge insurers from further liability under the contract. Instead, a warranty will be construed as a 'suspensive condition', meaning that insurers will not be liable in respect of losses occurring or attributable to something happening during the period of breach. Where a loss occurs when an insured is not in compliance with a term that 'tends to reduce the risk' of loss, the insurer will not be able to rely

on that non-compliance to exclude, limit or discharge its liability if the insured can show that its non-compliance did not in fact increase the risk of the loss that occurred in all the circumstances.

- In respect of fraudulent claims, the insurer will not be liable to pay any part of a fraudulent claim and may recover any money paid in respect of that claim prior to discovery of the fraud. The insurer may give the insured notice that the contract is terminated from the date of the fraud. The insurer can then keep the premium and will have no liability for claims arising after the fraud.

Hedging and derivatives

Catherine Jago
CJH Energy Limited

A ship-owning client tells its lawyer a sad tale of woe, the result of a claim amounting to US$3.5 million made against it by its charterer/cargo owner. The claim stems from a 1 million- barrel cargo of crude oil which the ship owner delivered 10 days late, due to a fault on the ship. How can the lawyer seek to reduce its client's financial liability? One way is to consider whether the claim is actually a result of market movement. If so, the cargo owner perhaps should have hedged (or did hedge), as a result of which most of the claimed loss would have been compensated for by hedging profits.

1. Introduction

Most oil companies, but not all, hedge their market price risk. In many cases in which the author is involved as an oil expert, the question arises whether quantum should be determined on the assumption that all companies hedge their price risk.

If the starting point is that hedging should be taken into account when quantifying a claim, either as a result of causation, or as a form of mitigation, is it then up to the claimant to show why it did not hedge its price risk or, if it did, to prove why it still has a claim.

Alternatively, should hedging be ignored when quantifying claims on cargoes of oil? Doing this could result in claimants receiving more money than they actually lost as a result of the action of the respondent or, alternatively, failing to claim all their losses. If the claimant has, arguably, unreasonably failed to hedge, is it possible to determine what profit or loss it might have made had if it had hedged the price risk?

Perhaps companies should carefully consider specifying whether hedging losses or profits can be included in any claims in their contractual terms, bearing in mind that hedging can work for or against a claimant. Contractual terms are discussed in the chapter on 'Damages'.

This chapter attempts to address some of these questions by discussing the broad principles of managing oil market price risk using hedging and how it is relevant to claims made for cargoes of oil that have either been lost, delayed or contaminated. It will show, by reference to some specific examples, how hedging can affect quantum and demonstrate that a claim can be reduced, negated or created by taking hedging into account. The chapter goes on to discuss some hedging tools and what makes one tool a more effective or more appropriate tool for hedging a particular market price risk. It then discusses why a trader may choose not to hedge and the risks he is taking by failing to hedge, as well as the risks a trader may still face where he does choose to hedge, which can result in losses not being covered or even created. Cases where hedging is not appropriate will also be considered, and examples will be given of where quantum has been constructed in such a way as to avoid having to take market loss, and therefore hedging, into account.

Needless to say, this is a huge subject, the surface of which the author can only hope to skim over in one chapter.[1] Further, this chapter is written from the perspective of an oil trader, not that of a lawyer, and so does not attempt to express a view on legal principles.

2. Principles of hedging

2.1 Introduction

Before considering hedging issues, it is first necessary to consider how the contractual price of oil is calculated.

The price of most oil traded in the international market is now rarely set at a fixed price; rather, it is linked to the price of a benchmark crude quoted by a daily oil price publisher such as Platts.[2] The benchmark prices are traded at a fixed price, in some cases determined through a transparent price discovery process known as the Platts Market on Close (MOC), whereby approved buyers and sellers transact electronically on set terms for a given commodity during a short set period of time, often referred to as the Platts Window.

The pricing formula for most crude oil and oil product transactions will generally specify a differential to a benchmark price, such as Dated Brent,[3] and will specify which published prices will apply. For example, the pricing formula may say that the price of a given crude will be 30 cents per barrel above the average Dated Brent quotation on the five Platts publication dates after Bill of Lading (BL). If the BL falls on Monday, March 2 2015, then the Dated Brent prices quoted by Platts on the five publishing dates after that date will be March 3, 4, 5, 6, and 9 (March 7 and 8 being the weekend, when Platts does not publish). The pricing formula will also specify which days will apply if the BL falls on a weekend or bank holiday. This may take the form of:

- a split weekend, using three days before the weekend and two days after, or two/three days; or
- taking the Friday quote twice if the BL falls on a Saturday; or
- taking the Monday quote twice if the BL falls on a Sunday; or
- whatever combination of published prices the two parties may agree.

The invoice price of this cargo of crude oil will therefore be US$50.605 plus 30 cents per barrel, which equates to US$50.905 per barrel.

---

1 For more information, readers should contact an oil expert or refer to the booklist at the end of this chapter.

2 Platts is part of McGraw-Hill and is a global provider of energy information and a source of benchmark price assessments.

3 Brent is a blend of several crudes from various fields in the North Sea. Its price is said to influence the price of as much as 95% of the crude oil traded worldwide.

Table 1: Example of Platts Prices (illustrative)

Date	**Low**	**High**	**Mean**
03-Mar-15	50.25	50.35	50.3
04-Mar-15	50.4	50.45	50.425
05-Mar-15	50.45	50.5	50.475
06-Mar-15	50.4	50.45	50.425
09-Mar-15	51.35	51.45	51.4
		Mean	50.605
		Differential	0.3
$/bbl		Invoice price	50.905

Table 2: Example of Platts Prices (illustrative)

Date	**Low**	**High**	**Mean**
24-Mar-15	55.35	55.4	55.375
25-Mar-15	55.25	55.35	55.3
26-Mar-15	55.1	55.2	55.15
27-Mar-15	53.4	53.45	53.425
30-Mar-15	54.05	54.15	54.1
		Mean	54.67
		Differential	0.5
$/bbl		Invoice price	55.17

However, what if, instead of loading and having a BL date of March 2, the cargo is loaded later? This may be because the ship's tanks are not clean or the owners/charterers choose to load another cargo first or the ship breaks down because it has not been maintained properly and a substitute ship has to be found. The new BL date is now March 23 2015. In the meantime, the market has risen. The charterer has lost its original purchasing opportunity and has had to renegotiate to purchase a later availability. It has been able to conclude a replacement purchase but has had to pay a differential of plus 50 cents per barrel, 20 cents above the original price, with pricing dates still five days after BL. The price is now calculated as follows.

The charterer wants to claim for the fact that, rather than paying US$50.905 per barrel for the oil, he has had to pay US$55.17 per barrel. He therefore wants to claim US$4.265 per barrel on an 800,000 barrel cargo, amounting to a claim for US$3,412,000.

The first thing to consider is whether the charterer's sale price has been altered by the delayed BL date. The cargo may now be worth more in the market, but can the charterer get more by delivering it into the sale contract he had committed to previously, assuming that he can still satisfy that sale?

It may be that the sale price is based on the average Dated Brent price of the month of loading, in which case the sale price is unaffected. The charterer has had to pay more for the cargo, but the sale price is essentially fixed. So long as the BL date is still in March, the sale price will be unaffected by exactly when the cargo loads, and so the charterer cannot get a higher price for the cargo, even thought he may have paid more for it. So has he lost US$ 4.265 per barrel? Perhaps not, if he has hedged.

If the charterer had hedged, he should have done so when the breach occurred and the ship failed to load. When the resale actually priced, he would reverse his hedge. In so doing, he could try to protect himself from the market movement element of his price. In this case, he could potentially protect himself from the market rising by US$4.065 per barrel (the difference between US$50.905 and US$54.67). What he could not necessarily protect himself from is having to pay an extra 20 cents per barrel differential. His claim should therefore be for 20 cents per barrel, not for US$4.265 per barrel. On a cargo of 800,000 barrels, this could save the lawyer's shipowner client US$3,252,000.

What if the sale contract the charterer already had, and which he was still able to perform, was also based on prices current around BL date? Let us assume that his sale contract was based on prices as at 5 days after BL but at a differential of 60 cents per barrel. The sale price should have been US$51.205 (US$50.605 plus 0.60), but is now US$55.27 (US$55.67 plus 0.60), so he has made an extra US$4.065 per barrel on his sale. His crude oil purchase has cost him an extra US$4.265 per barrel, so he has lost US$0.20 per barrel. This is because he has had to pay an extra 20 cents per barrel for the replacement oil. The charterer had executed a back-to-back transaction. He had no market price risk because he was buying on five days after BL and selling on 5 days after BL. So long as he could still do this, he lost nothing due to the market movement, but he had to pay a higher market premium for the oil from the replacement provider. Effectively, he was hedged by having matched his purchase and sale pricing basis (a back-to-back transaction); he therefore had no underlying market price risk.

Returning to the unfortunate charterer with the absconding ship, what if, instead of doing a back-to-back transaction where the pricing basis was exactly the same, he did it on a slightly different pricing basis, say 3 days after BL? He was at risk of market movement between the average of the five days after BL and the average of the three days after BL in any event, regardless of the delay. Perhaps he should have protected himself anyway from this contractual price risk by hedging. If he had done so, he could also have protected himself from the loss he might have made from the delay.

In this scenario, the charterer would in fact have made a gain on the physical purchase and sale by the cargo loading later and pricing being delayed. His sale price would have been US$50.40 plus 60 cents or US$ 51, but now he achieved US$55.275 plus 60 cents or US$55.875, an extra US$4.875 per barrel. So, while he may have paid an extra US$4.265 per barrel, he would have received an extra US$4.875 from his sale because of the way the market has moved.

Table 3: Examples of Platts Prices (illustrative)

Date	Low	High	Mean
03-Mar-15	50.25	50.35	50.3
04-Mar-15	50.4	50.45	50.425
05-Mar-15	50.45	50.5	50.475
		Mean	50.4
		Differential	0.6
$/bbl		Invoice price	51
Date	**Low**	**High**	**Mean**
24-Mar-15	55.35	55.4	55.375
25-Mar-15	55.25	55.35	55.3
26-Mar-15	55.1	55.2	55.15
		Mean	55.275
		Differential	0.6
$/bbl		Invoice price	55.875

If the charterer had hedged, he would have lost out on this gain and possibly lost more than he gained on the physical. He might conclude that although he made a gain on the physical overall, his hedge meant that he made a loss, which would be the basis of his claim. This point will be discussed later by reference to some examples.

2.2 Why is there a need to hedge?

There can be no doubt that the oil market is volatile. Late 2014 saw the oil price plummet from the dizzy heights of above US$100 per barrel to below US$50 per barrel within a few weeks.

Crude oil spot prices 1990–2014

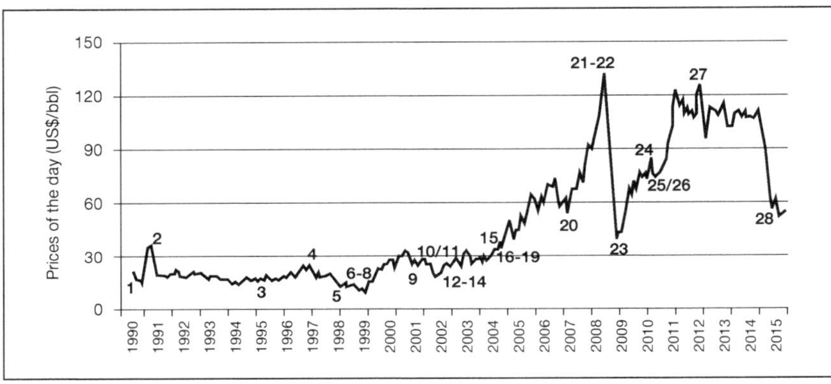

Source: EIA – www.eia.gov (crude oil spot prices)

Key:

1. 1990 (Aug) – Iraq invades Kuwait, UN embargo on Iraq
2. 1991 (Nov) – Soviet Union collapses
3. 1995 – UN resolution to allow partial resumption of Iraq oil exports in 'oil for food' deal
4. 1997 – Kyoto Agreement proposed to limit greenhouse gases
5. 1998 – Asian economic crisis
6. 1999 – US sanctions against Libya lifted
7. 1999 – Panama Canal reverts to Panamanian sovereignty
8. 1999 – Y2K fear
9. 2001 (Sep 11) – Terrorist attacks on the United States
10. 2002 – US threatens invasion of Iraq
11. 2002 (Nov) – UN weapons inspectors return to Iraq
12. 2003 (Mar 19–20) – US-led invasion of Iraq begins
13. 2003 (Apr 9) – US takes Baghdad
14. 2003 (Aug 14) – Electrical failure causes blackout in New York State and Ontario
15. 2004 (Nov) – George W Bush re-elected US President
16. 2005 (Jul 7) – Terrorist attacks in London
17. 2005 (Jul 24) – Iran and Iraq sign an oil trading agreement
18. 2005 (Aug 29) – Hurricane Katrina strikes the Gulf Coast of the US

19. 2005 (Sept 23) – Hurricane Rita strikes Gulf Coast
20. 2007 (Mar) – European Union introduces new environmental regulations to reduce greenhouse gas emissions
21. 2008 (Sept–Nov) – Global recession fears
22. 2008 (Nov–Jun 2009) Global recession a reality
23. 2009 (Dec 7–20) - Climate change conference in Copenhagen, Denmark
24. 2010 (Apr 15) – Eyjafjallajokull volcano in Iceland erupts
25. 2011 (Jan–Apr) – Unrest in various countries of the Arab world ('Arab Spring')
26. 2011 (Mar 11) – Earthquake and tsunami hit Japan and cause considerable damage to the Fukushima nuclear plant. Surge in demand for fuel oil
27. 2012 (Jul 31) – Major electrical power failure in India
28. 2014 – Opec fails to cut back production. Surge in non-Opec supply

Source: Geohelp – www.geohelp.net (world events)

Any trader with a fixed price purchase for which he did not have a fixed price sale would have been in big trouble unless he had protected his position by hedging. It is not surprising to have seen such a huge rise in interest in hedging oil prices. One tool, the ICE Brent contract, now trades about five times the total world oil consumption every day – and that is just one contract for one futures market.

ICE Brent Futures and global oil consumption

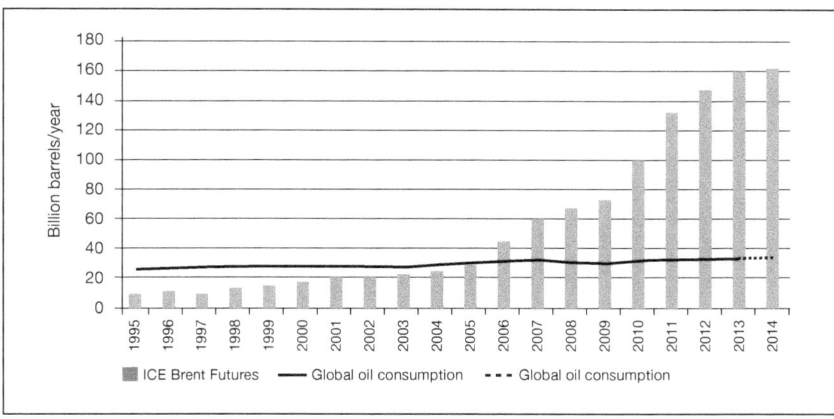

Sources: BP – www.bp.com (global oil consumption); Bloomberg – www.bloomberg.com (2014 global oil consumption forecast); The ICE – www.theice.com (hedging numbers).

So, what does the magic saviour of hedging involve?

If a trader hedges, he takes an equal and opposite position in the hedge to his physical position. The author prefers to think about it in the following way. If the trader owns some oil, which has already priced but his sale price has not been fixed (because either he has not yet sold the oil or the pricing formula of his sale has not started to establish), then he is fearful that prices will fall before his sale price is fixed.

Brent prices (nominal)

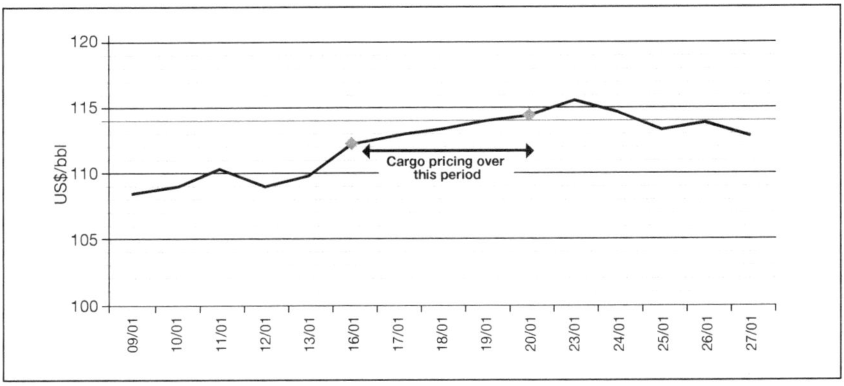

What the trader needs is a tool that will make him money if the market does fall. Then, even if he loses money on his physical sale, he will make money on his hedge to offset that loss. He therefore needs to sell a hedge. When his sale starts to price, such that instead of his sale being a floating price it becomes a fixed price, he can then buy back his hedge, at a lower price if the market has fallen, making money on the hedge which can be used to compensate for his lower physical oil sale price.

Brent prices (nominal)

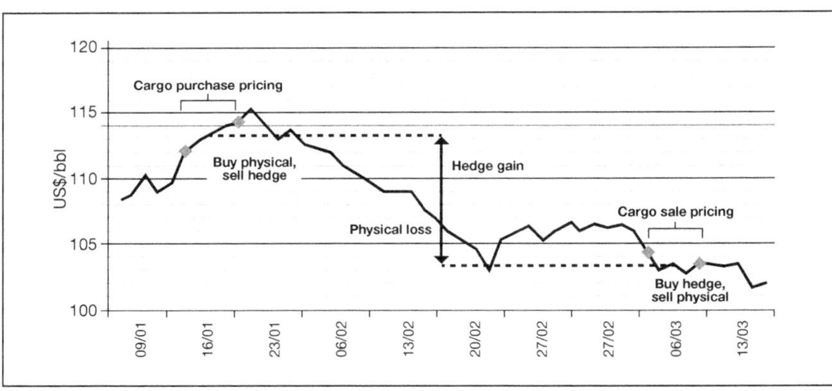

Let us take a very simple example. A trader has purchased a cargo at Dated Brent (Dtd Brent) plus 40 cents priced on Day 1 and sold it at Dated Brent plus 60 cents to be priced on Day 10.[4] He wants to protect his 20 cents a barrel profit, regardless of how the Dtd Brent price moves.

4 For simplicity, the cargo is priced on one day. In reality, this is unusual.

Example 1: Market falls

	Physical	Hedge
Day 1	Purchase Dtd Brent plus 40 cents at US$50.40	Sell Dtd Brent at US$50
Day 10	Sell Dtd Brent plus 60 cents at US$48.60	Buy Dtd Brent at US$48
P/L	–1.80	2.00

P/L without hedge US$ –1.80 per barrel
P/L with hedge US$ 0.20 per barrel

The trader, his manager, the shareholders, the banks and his family are all happy as he protects his identified profit (and his bonus).

However, what if, instead of the market falling, (as the trader feared), the market rose.

Example 2: Market rises

	Physical	Hedge
Day 1	Purchase Dtd Brent plus 40 cents at US$50.40	Sell Dtd Brent at US$50
Day 10	Sell Dtd Brent plus 60 cents at US$52.60	Buy Dtd Brent at US$52
P/L	2.20	–2.00

P/L without hedge US$ 2.20 per barrel
P/L with hedge US$ 0.20 per barrel

The trader, his manager, the bank and his family are all happy – the trader has secured the profit he originally intended of 20 cents per barrel.

His shareholders, on the other hand, may ask, "Why did you hedge? Without the hedge you would have made more money. This shows that hedging is wrong."

While the second statement – that without the hedge the trader would have made more money – is perfectly true, of course this ignores the point of hedging, which is to protect oneself from the market risks one fears might happen, not to make the most money. Hedging replaces uncertainty with a degree of certainty.[5] It locks in the trader's perceived profit. The trader identified that there was a 20 cents margin in the deal for him and he wanted to secure it. His reason for hedging was to protect that margin, not to get the best price. Hedging reduces future oil price uncertainty.

In cases litigated, the purposes of hedging is often attributed to mitigation after a breach has been established – that is, an attempt made by the claimant to minimise his losses. On the other hand, hedging losses could be linked to causation – did the breach cause the hedging loss? This is not, however, a chapter on the law, so whether in any particular case hedging can be considered to be mitigation or causation will not be considered here. Either way, hedging – or not hedging – can have a major impact on the overall amount a claimant may lose as a result of a breach.

5 Not absolute certainty because of basis risk.

A line needs to be drawn, however, when determining whether claims flowing from hedging losses or gains have arisen as a result of managing price risk or from speculation. The two can be closely linked, often using the same tools.

Hedging is designed to reduce risk and is used by those companies that are more risk-averse. Speculation, by contrast, generally increases risk and is used by companies wanting to take on risk for the rewards it may bring. An example of speculation is the buying of futures contracts with no underlying physical position. Alternatively a trader may take a view that the market is going to decline, in which case he may sell a hedge, and when it moves down he can buy it back at a profit. If, however, the market moves up, the trader will then lose money. Speculation exposes him to market movements,[6] whereas hedging allows him to seek to minimise his exposure to market movements.

A hedge taken out to protect a market price risk can become speculative if it is not managed carefully. For example, if the hedge is not closed out correctly, the trader could be over- or under-hedged compared to his physical exposure, so that any losses on his hedge are not compensated for by a gain in his physical price or vice versa. A 50-kt cargo may be pricing over five days. If the trader lifts the whole of his hedge on day one, he is exposed to market movements for 80% of his cargo, or 40kt. If the market moves against him, he could lose a lot of money. Alternatively, he could make a lot, but he is speculating on market movements, not hedging his market risk. As a result, is it correct for losses associated with this speculative position to be included in any claim? The question to ask is, were these losses incurred as a result of the breach or as a result of the trader speculating?

3. Hedging tools

This chapter can only touch upon the types of tool that form the arsenal available to a trader to protect himself from market price risk.[7]

The simplest hedging tool is probably the futures market. A trader can buy, or sell, a set volume per lot of a defined oil commodity for delivery[8] during a forward month. Futures contracts are traded anonymously through a regulated exchange, so trading is not bipartite. Financial and performance security is guaranteed through a clearing mechanism and rigid contract specifications.

Financial performance is facilitated through the process of margin calls. Initial margins have to be paid if a trader wishes to hold open positions overnight, or, in other words, hedges that have not been closed out – a purchase that has not been matched with a sale or vice versa. In addition, if at the end of the trading day the hedge is losing money (ie, the settlement price is below the original purchase price or above the original sale price), then variation margins have to be paid to cover the difference between the price at which the position was taken out and the closing price of that trading day. If, on the other hand, the hedge is making money, then a

---

6 Assuming that he is not using options or spreads – options allow the buyer to restrict his maximum exposure to the premium paid for the option. Spreads trade the difference between two commodities, which reduces the market risks by comparison with outright speculative positions.

7 For more information, see the recommended reading listed at the end of this chapter.

8 In fact, many hedging tools are financially settled and no physical delivery takes place.

positive variation margin is transferred to the trader's hedge account. The benefit of this process is that the trader has financial security. The disadvantage is that a profitable physical position may well be hedged with a futures or swaps position that is losing money. The profit on the physical cannot be realised until payment is made in accordance with the terms of the contract, whereas the variation margin money has to be paid up immediately.[9]

The most active futures contracts are traded on two exchanges – the Intercontinental Exchange (ICE) and the New York Mercantile Exchange (NYMEX), which is part of the Chicago Mercantile Exchange (CME) group. The range of futures contracts is very limited, which means that hedging using futures contracts can be far from perfect and can introduce Basis Risk.[10] Partly in response to the limited range of futures contracts, additional Over The Counter[11] hedges, known as swaps, have been created.

Swaps are available in a far wider range of crude oil and product markets and locations than futures. They are essentially a futures market but involve swapping a future floating price for a fixed price today – for example, the average Platts price for high sulphur fuel oil FOB ('free on board') Singapore over the month of March with a fixed price of US$400 per mt. No delivery of oil takes place and settlement is purely financial. Originally, swaps contracts were not financially secured, but they are now increasingly regulated and cleared by Clearing Banks offering financial security to traders, their banks and their shareholders and assisted by guidance from bodies such as the International Swaps Dealers Association. The downside of having these cleared is that margin calls have to be paid as in futures contracts, which can deplete a trader's credit lines.

Swaps contracts are used extensively for hedging crude oil, such as a hedge known as Contracts For Difference for Brent, which allows the trader to swap the floating price differential between Dated Brent[12] and the 25-day BFOE[13] forward contract for a fixed differential.

Other tools available to the trader are forward markets. Historically, these were Over The Counter contracts for future delivery of whole cargoes concluded directly with a counterparty. In some markets, such as Brent and Naphtha, they continue to flourish, but many have now been replaced with active swaps markets, such as for jet and fuel oil.

Finally, traders can use options. Unlike futures, swaps and forward contracts, options give the buyer of the option the right but not the obligation to buy or sell the commodity at a fixed price for a date in the future.

Options tend to be used more for strategic rather than operational hedging

9 Actually, usually the following morning.

10 The difference in the way the hedge moves compared to the hedged physical position.

11 Two parties negotiate directly without using a standardised product traded on regulated exchanges. They involve bipartite contracts negotiated between two parties without using any central organisation and with each party knowing who their counterparty is. Each is exposed to the other for performance and payment.

12 Physical cargoes of Brent Blend which have designated dates for loading at Sullom Voe terminal up to 25 days forward.

13 An OTC forward contract for cargoes of Brent, Forties, Oseberg or Ekofisk loading in the future.

purposes. For example a crude oil producer may buy a put option,[14] which gives him the right to sell his crude oil at a given price several years into the future. On the basis of the option, he may be able to secure funding. Operational hedging, on the other hand, aims to manage the market price risk associated with a single cargo, or perhaps a few cargoes, not the volumes to be produced from a field for 10 years hence.

When assessing the relevance of hedging to legal claims, what is generally being considered is operational hedging, which tends to look to the shorter term.

Given such a panoply of hedges, how does the trader select which tool to use? The most important thing for a trader to consider when undertaking hedging (as opposed to speculation) is to match the hedge as closely as possible with the physical position and associated price risk. In order to do this, the trader needs to look at what his risk is.

Let us take an example of a trader who has loaded a cargo of 90,000 mt of fuel oil in Europe and is sailing it to Singapore. He has yet to sell the cargo but is confident he will do so as there is a huge bunkering market for fuel oil in Singapore. His purchase price was based on Mediterranean fuel oil prices, but his market risk exists not in the way Mediterranean prices might move any more, but in the way Singapore prices move, as it is these prices which will influence the price he will ultimately be able to achieve when he sells his physical cargo.

The trader therefore needs ideally to hedge his risk in Singapore fuel oil. There is, however, no futures or active forwards market in Singapore, but there is a swaps market there. Is that market active enough for him to hedge the whole cargo in one go, however? There is a more active market in north-west Europe, so perhaps the trader will initially hedge his risk in that area's fuel oil market. The market in north-west Europe may not move in tandem with market movement in Singapore, so once he has established his hedge as quickly as possible in Europe, he may then transfer his position, with more leisure as he has secured some price protection, to Singapore by using an East-West fuel oil swap which converts his hedge from Europe to Singapore. This is not, however, very active with regard to the quality of the oil he has. Thus, he does the East-West swap in, say, high sulphur 180 centistokes fuel oil and then does another swap in the 'Visco Spread'[15] market to achieve his hedge in the market which most closely reflects his risk.

Other dilemmas he may have to consider are the following.

(a) *Volumes*

Unfortunately, the swaps do not trade in exactly the volume of the cargo – the trader actually loads 91,542 mt. So, does he over-hedge slightly or under-hedge?

(b) *Timing*

The cargo is due to discharge in December but swaps for December will only be trading for part of that month, so the trader chooses to take out his hedge in January

---

14 The buyer of a put option has the right but not the obligation to sell a commodity. The buyer of a call option has the right but not the obligation to buy a commodity.

15 Viscosity, measured in centistokes, is one key specification for fuel oil.

swaps. Will January contracts respond to market forces in the same way as the underlying physical market for December pricing responds?

In the case of a breach, timing of the hedge can become critical and can be the cause of additional losses. Let us take as an example the situation where delivery of the cargo is delayed by three months due to documentary problems. The trader loses the sale he secured during the journey and is now exposed to Singapore prices again. By the time his cargo is delivered, he is forced to sell at a lower differential than that at which he originally sold. The good news, however, is that the underlying market has risen, so that he actually sells the cargo at a higher absolute price than he had originally expected. As a result of the way the market has moved, he has made money as a result of the breach and so, on the face of it, has no claim.

The trader has, however, lost money on his hedge because he protected himself from the fear of falling prices by selling a hedge. Contrary to his fears, though, the market has risen, so he loses money on the hedge. Furthermore, because the cargo was delayed by three months, he had to close his original hedge position and reopen it in a later contract month. It is no good holding a hedge in January contracts if, by the time the cargo actually prices, January contracts are no longer trading. The trader therefore has had to roll his hedge in a backwardated[16] market. This means that every time he bought back the original hedge and sold it in the forward month in order to maintain his hedge through the delay, he lost more money because he was buying back at a higher price than he could sell.

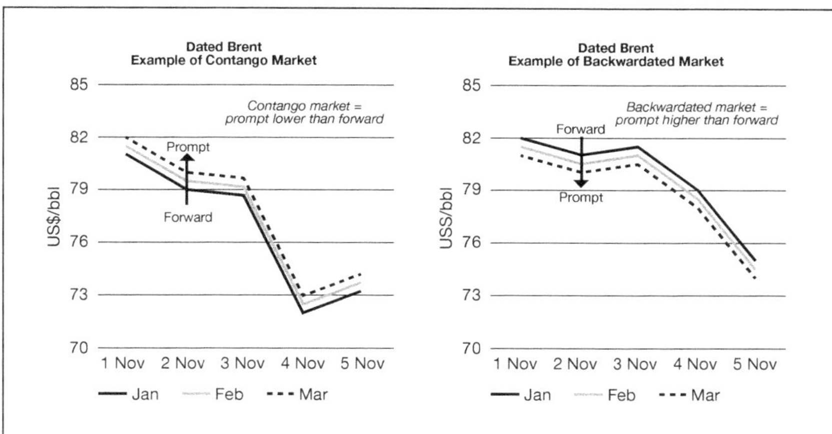

The result in this example may be that a claimant may make money on the physical due to the breach, but lose more money on the hedge than he made on the physical. In other words, quantum comprises hedging losses, not losses on the price of the physical oil.

---

16 A backwardated market is one where, on a particular day, prices for prompt oil are higher than those for oil being delivered in the future. A contango market is where prompt oil is worth more than oil for delivery in the future.

The above example shows both how Basis Risk can affect the choice of hedge used and Basis Risk can result in a far from perfect hedge. Hedging cannot always protect a trader from losses due to market movements. Sometimes the Basis Risk is so large that it is questionable whether the hedge is worth opening. This leads to discussion of which tools to select for successful hedging, and when it is better not to hedge at all.

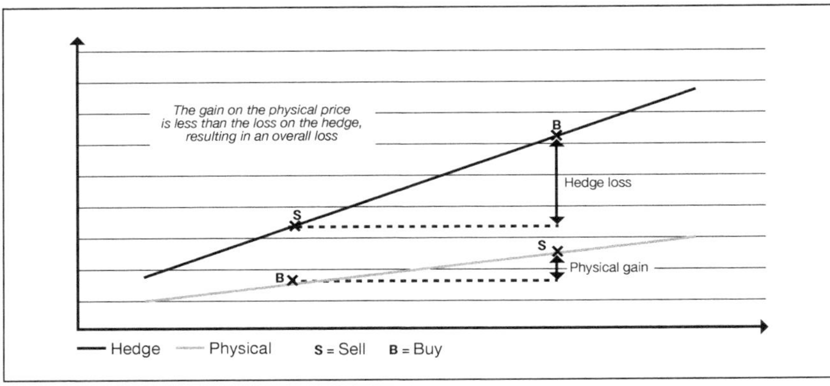

4. How to select tools

Traders select their tools, like any good tradesman, in accordance with their needs and the risks they are prepared to take. One trader may decide that he is content to hedge his fuel oil cargo using crude oil hedges such as ICE Brent, even though there is a quality basis risk, because the liquidity is good in Brent futures and not so good in fuel oil swaps. He thinks that the correlation between the two is sufficient for his needs.

Another trader, however, may decide that he is prepared to hedge 50% of his risk and accept the market risk of the remaining 50%. It could be said that he is managing the risk for half his portfolio but speculating that the market will move in his favour on the remaining unhedged portion of his trading exposure.

Another trader may decide that he will take out a swap in the locational market which best matches where his risk is, thereby reducing his basis risk. As time passes, the market moves more in his favour and he is prepared to close out his hedge, which of course is now loss-making as his physical is making him money, and is content to ride the favourable wave on his physical completely exposed to market movements. He may be fortunate and make a lot of money. He may, however, be less fortunate and lose on his physical. The market could suddenly turn and he will be unable to re-establish his hedge quickly enough to protect himself from losing money on his physical. It could be said that he is speculating rather than hedging. Should any losses he incurs as a result of not having fully hedged, or even not having hedged at all, be validly included in quantum?

There are certain times when not hedging may be sensible, or at least reasonable. One example may be if the only tools available have poor correlation with the market risk the trader is facing. For some more unusual oil products, it may be particularly difficult to find a hedge with good correlation, such as some feedstocks including vacuum gasoil, or blendstocks such as reformate. There may be ways to hedge these products using a combination of hedging tools, but some traders may say they would prefer to take the risk. Alternatively,

- perhaps the risk will only exist for a very short period, say over the weekend; or
- the volume of the physical is relatively small, say 2,000 mt.

It is said that certain companies do not hedge. For example, Exxon have often been exemplified as a major oil company that chooses not to hedge its market price risk. The reason postulated is that it is so large, and vertically integrated from the wellhead to the pump, that it is effectively self-hedged. Other companies, or perhaps more correctly, cultures, refuse to hedge because they consider hedging to be a form of betting. It is, however, the case that most traders hedge their price risk, or at least consider hedging, and then take a view whether to hedge on the basis of some of the criteria previously discussed.

Of course, if a trader is never exposed to market risk, he will not have to consider hedging. It could be argued that a trader is never unexposed to market risk but, leaving that conundrum to one side, if a trader buys and sells on exactly the same terms, and the same pricing formula, except for the differential, then he may have no need to hedge. The requirement to hedge or not to hedge may, however, be changed if some event intervenes to disconnect the trader's back-to-back transactions.

This chapter has examined examples of where the breach results in a delay of loading or discharging the cargo. What about if a cargo becomes contaminated?

A cargo of gasoil loads in northern Europe and is due to be delivered into the Mediterranean. When it reaches its intended discharge location, however, one of the key parameters, such as pour point, is off-specification as a result of contamination from the ship's lines at the loading port. The cargo owner's buyer refuses to accept the off-specification oil. The cargo owner, therefore, either has to find another buyer for that oil or to treat the oil in order to correct the pour and then resell it.

Let us assume that the cargo owner tries unsuccessfully to correct the off-specification oil and so eventually decides to sell it as unsound oil. Not only does he have to discount the oil because it is unsound but, in the meantime, the market has also fallen and so the absolute price he is able to achieve is lower. The cargo owner's claim is made up of the difference between the price at which he was going to sell the oil, say $US550 per mt, and the price at which he actually sells it, say US$425 per mt. Only US$25 per mt of this loss is, however, due to the contamination in the gasoil. The remainder is due to the fact that during the time the cargo owner tried to correct the contamination and before the final sale actually priced, the market fell US$100 per mt. Should the shipowner, if held liable, pay the cargo owner US$125 per mt when his breach, which caused contamination of the original cargo, only caused US$25 per mt of the loss? The remainder could be hedged, but how much could that have saved?

Instead of looking at the absolute prices of the sound and unsound cargoes at two different dates, another way of determining quantum is to look at the prices the sound and unsound cargoes were sold for against a benchmark. In this way, the market movement is taken out of the equation.

Thus, in the example given above, if the original sale was agreed at the average of Platts Mediterranean Gasoil quotation five days after NOR ('notice of readiness') plus US$6 per mt, but the resale of the now unsound cargoes was concluded at the average of Platts Mediterranean Gasoil quotation five days after NOR less US$19 per mt, the loss caused by the contamination can be measured as US$25, or the difference between plus US$6 and minus US$19 per mt. In calculating the loss in this way, any issues around market movement, and therefore whether the cargo owner did or did not hedge and if so, how much he did or did not make on the hedge, can be taken out.

The examples given so far may look relatively straightforward and one might think that it would be easy to measure the extent of any hedging losses or gains for a specific trade in order to determine quantum correctly by looking at the hedges the trader has undertaken. Sadly, however, real life is often not that simple. The problem is that many traders do not hedge each cargo individually. Trying to extract from their whole hedging portfolio the hedges specifically attributable to the market risk associated with a single cargo can be worse than looking for the proverbial needle in a haystack.

Traders will often hedge their whole book. Thus, if he was long[17] crude oil in one market but short in the other, the trader may conclude that the two positions balance each other out, and will not hedge at all. He is self-hedged by his book positions, when looked at holistically, in that overall he is not exposed to market movement. He is indifferent to whether the market moves up or down. By not hedging, he saves brokers' commissions, initial margins, variation margins and general calls on his credit lines.

The long position may, however, close out (ie, may price) before the short, at which stage the trader's position is now short, so that he has to hedge this new market risk. Multiply this by the many cargoes a trader may have at any one time, and then by the numbers of traders that may be in an organisation, and it is clear that extracting the exact hedges used to hedge a single cargo individually may be impossible – or, at the very least, a huge logistical exercise that a trader may be extremely reluctant to spend his time on, preferring to be out in the market making more money. This is when an expert may be required to construct a hedging strategy that simulates what the trader might have done had he hedged this single cargo, and the price risk associated with the specific cargo to which any litigation relates. Models can be created which create a base hedging strategy and which can be sensitivity-tested for changes to that strategy, such as different tools, contracts, timing, or hedge volumes.

5. Conclusion

From the above very brief analysis it can be seen that hedging can dramatically affect quantum in a case involving commodities such as oil. It can:

- increase or decrease a claim;

---

17 'Long' means that one has bought or has committed to buy a commodity and is therefore potentially at risk to falling prices. 'Short' means that one has sold or has committed to sell a commodity and is therefore potentially at risk to rising prices.

- generate a claim where there appears to be no loss; or
- completely wipe out a claim where, on the face of it, a litigant suffered a loss.

To exclude hedging losses contractually from any claim may work in a claimant's favour, but can also work against them. Leaving it open to include hedging losses or gains may seem prudent, but identifying the actual hedging losses can create levels of complication in determining quantum. Either way, it is important to understand whether a claim can be challenged in respect of hedging losses or gains early on in the proceedings, as it may make the difference between settlement and court judgment. In particular, if attempts are to be made to reach some mediated settlement, it is particularly important to understand where hedging might play a role.

Recommended reading

Liz Bossley, *Trading Crude Oil – The Consilience Guide*

Catherine Jago & Liz Bossley, *Trading Refined Oil Products – The Consilience Guide*

Damages

Iain Sharp
Rodyk & Davidson LLP

1. Introduction

Under English law, damages for breach of contract is a common law remedy. The term 'damages' has been defined "quite simply as an award in money for a civil wrong"[1] that is intended to compensate an innocent party for the damage, loss or injury suffered through that breach. In line with this principle, a party that commits a breach of contract is liable to pay in damages a sum of money that would place the innocent counterparty in the same position as if the contract had been performed.

In cases involving the sale of goods (or, for that matter, the supply and carriage of goods), English statute and case law provides guidance as to how damages should be assessed. The right to reject goods and documents will customarily be accompanied by a right to claim damages, for example, for non-delivery by the seller. Further, where a breach of a contractual term regarding quality occurs, the buyer is entitled to bring an action against the seller for damages.[2]

As is often the case in international sale contract disputes, parties may agree that damages are to be assessed on the basis of market value. There can, however, be disagreement as to how and when that market value is to be determined.

This chapter discusses how damages are assessed in a number of common claim scenarios in international sale contracts, including non-delivery and late delivery of goods.

2. Causation and remoteness of damage

Two basic principles of English law feature in all cases in which damages for breach of contract are assessed. The principles – of causation and remoteness – restrict the type of loss that can be recovered by way of damages.

Causation is the process of proving that the other party has caused the innocent party's loss. In contractual disputes, there must be a causal link between the breach of contract and the resulting damage. The party in breach must have caused or contributed to the loss.

The traditional test for remoteness of damage is as set out in the decision of the Court of Exchequer in *Hadley v Baxendale*[3] and as clarified by the English Court of Appeal in *Victoria Laundry (Windsor) Ltd v Newman Industries Ltd*[4]:

---

1 *McGregor on Damages* (19th Edition), paragraph 1-001.
2 Section 53(3) of the Sale of Goods Act 1979.
3 (1854) 9 Exch 341.
4 [1949] 2 KB 528.

> *... [T]he damages which the other party out to receive in respect of such breach of contract should be such as may fairly and reasonably be considered either arising naturally ie, according to the usual course of things from such breach of contract itself or such as may reasonably be supposed to have been in contemplation of both parties at the time they made the contract as the probably result of the breach of it.*

The first limb of the *Hadley v Baxendale* test allows for the recovery of losses arising naturally from breach of contract.

The second limb allows for recovery of 'unusual' losses that do not flow naturally from the breach of contract, but rather as a result of special knowledge possessed by the party who breached the contract.

In *The Achilleas*,[5] which involved a claim by shipowners for the late redelivery by charterers of a ship let to them under a time charter, Lord Hoffman introduced a new criterion for the determination of remoteness of damage, namely whether or not the party in breach had assumed responsibility for the loss which had occurred as a result of the breach.

There has been much debate as to whether *The Achilleas* has changed the orthodox approach to remoteness of damage. It appears that the classic test for remoteness of loss will apply in the great majority of cases, and that the 'assumption of responsibility' test is likely to apply in unusual cases where the surrounding circumstances or general market expectations make it necessary to examine whether or not a party has assumed responsibility for losses of the kind suffered by the innocent party.[6]

3. Mitigation

Following a breach of contract, the innocent party has a duty to mitigate[7] the loss it has suffered. Failure to mitigate loss may prevent that party from recovering damages for avoidable loss. A standard defence which the defaulting party often invokes to reduce the damages payable is that the innocent party has failed to act reasonably to mitigate its loss. The burden of proving a failure to mitigate falls, however, on the defaulting party.

As one might imagine, the courts are generally sympathetic to efforts made by an innocent party seeking to deal with a breach of contract. The requirement to take 'reasonable steps' to mitigate loss is not a particularly high standard. This so-called 'duty' requires reasonable steps to be taken to limit the losses that are incurred (and also to avoid incurring unnecessary expenditure in seeking to remedy the breach). An innocent party need not, however, take unusual steps that would be outside the normal course of its business, or even incur undue costs. Reasonable costs of mitigation incurred by the innocent party will generally be recoverable from the defaulting party.

For example, in contracts for the sale of goods, the claimant seller will usually have to give credit for the market value of the goods at the time of termination. This requires the defendant buyer to establish two things:

---

5 *Transfield Shipping Inc v Mercator Shipping Inc* [2009] AC 61 (House of Lords).

6 See, for example, *Sylvia Shipping Co Ltd v Progress Bulk Carriers Ltd (The Sylvia)* [2010] 2 Lloyd's Rep 81.

7 This is not a 'duty', in the normal sense of the word, but a rule of causation.

- that there is a market for the goods; and
- the market value of the goods.

The question of how market value has been addressed by the courts when assessing damages is considered in section 8 of this chapter.

It should be noted that the courts are likely to penalise a claimant in damages where it has demonstrably and unreasonably failed to take any steps to mitigate its loss, or where the steps that it has taken are plainly inadequate. For example, where a claimant could potentially mitigate its loss by selling goods in the market, the courts will usually expect to see evidence of meaningful attempts to do so.

4. Non-delivery

The *prima facie* measure of damages for non-delivery of goods is prescribed by section 51 of the Sale of Goods Act 1979 (the 1979 Act), which states:

(1) Where the seller wrongfully neglects or refuses to deliver the goods to the buyer, the buyer may maintain an action against the seller for damages for non-delivery.

(2) The measure of damages is the estimated loss directly and naturally resulting, in the ordinary course of events, from the seller's breach of contract.

(3) Where there is an available market for the goods in question the measure of damages is prima facie *to be ascertained by the difference between the contract price and the market price or current price of the goods at the time or times when they ought to have been delivered or (if no time was fixed) at the time of the refusal to deliver.*

In establishing the existence of an 'available market', the essential requirement is that the buyer should be able to go out and purchase equivalent goods in the market. If the buyer can do this, the courts will award damages by reference to the price paid in the 'available market'. Occasionally, however, equivalent goods are either not immediately available or not available from an immediately proximate location. In such cases, the courts will look at whether it was reasonable for the buyer to wait or buy equivalent goods from further afield.

Damages for non-delivery of goods contracted for are assessed by the English courts by reference to the difference between the contract price and the value of the goods on the date when they should have been delivered to the claimant (and not by reference to the difference between the contract price and the date of the defendant's repudiation or the date the claimant accepted the defendant's repudiation as bringing the contract to an end).

In cases of non-delivery of goods, claims for consequential losses are less likely to arise where there is an available market in which the buyer can purchase equivalent goods for its use or resale. In the absence of an available market, claims for consequential losses can be more frequent. Consideration will then need to be given to contractual limitation and/or exclusion clauses (as to which, see section 13 of this chapter).

Pursuant to section 54 of the 1979 Act, the buyer has a right to recover interest and consequential losses, such as additional costs or expenses incurred as a result of the seller's breach. This may include compensation paid to a sub-buyer – something which is not uncommon where a chain of contracts is involved. Loss of profit under a resale contract may also be recoverable in limited circumstances, namely where:

- the seller should have contemplated, at the time the contract was made, that the buyer was (or probably was) buying the goods for resale; and
- the buyer could fulfil its obligations under the resale contract only by delivering the same goods.

5. The measure of damages for non-delivery

Glencore Energy UK Ltd v Transworld Oil Ltd[8] involved the FOB ('free on board') sale of a cargo of Nigerian crude oil (known as Ukpokiti) and a claim by the buyer (Glencore) for damages against the seller (Transworld) for non-delivery. In that case, while the tanker was waiting to berth, the crew of a tug tasked to assist in that operation was kidnapped. Although the kidnap was resolved relatively quickly, and Transworld claimed that the tanker could have lifted the cargo, it had sailed away by then. The parties then sought to put in place alternative arrangements, but these ended in failure. In broad terms, the issues that arose for decision were:

- whether the contract came to an end, or was affirmed; and, if so,
- whether the claim was nevertheless time-barred under the NNPC (Nigerian National Petroleum Corporation) terms; and if not,
- what was the measure of damages for breach?

It was common ground that there was no available market for Ukpokiti oil, so that the measure of damages was to be assessed under section 51(2) of the 1979 Act. Glencore claimed the difference between the contract price and the value of the oil on the date when it ought to have been delivered. The key issues on quantum were:

- the correct approach to calculating damages under section 51(2) in the circumstances of the case; and
- whether the reduced hedging loss made by Glencore in closing out early its position following Transworld's breach was to be taken into account as mitigation of loss for the purpose of assessing damages.

On calculation, Glencore claimed the difference between the contract price and the value of the oil when it ought to have been delivered. Further, the value of the oil at that time would most probably have been the average of five Platts quotations following the bill of lading, this (Glencore submitted) being a typical pricing clause in the trade.[9] Its damages calculation was US$11,112,626.

Transworld's case at trial was that the market price of the oil was to be assessed by reference to the Dated Brent swaps price prevailing some weeks earlier, that being the time when a contract for delivery of a cargo of Ukpokiti oil in late June would typically have been made. The swaps price formula, it was submitted, properly reflected how a trader like Glencore would in fact have executed such a transaction. That formula provided a recoverable loss of US$7,562,797.

The court agreed with Glencore's approach. In summary it held that:

---

8 [2010] EWHC 141 (Comm).

9 This was the formula used in the sales by Glencore to BP and to Morgan Stanley. As to valuation by reference to Platts prices, see section 8 of this chapter.

- it would be wrong in principle to adopt (for the purposes of damages) a time equivalent to the time of the making of the contract (the 'swaps formula') rather than its performance (the 'futures formula');
- Transworld's formula for measuring market price was impossible to apply on the basis of any certainty;[10] and
- the price of Dated Brent was, by its nature, a price for future delivery: it provided a ready means for calculating the market price of the oil in question at the time of non-delivery.

6. Mitigation, hedging and the impact on damages

On mitigation in the *Glencore v Transworld* case, Glencore argued that the reduced hedging loss should not be taken into account, because the only steps it could have taken in mitigation was to purchase a replacement physical cargo, which was not possible as there was no available market. The hedges were independent transactions (akin to an insurance policy) and, as such, should not be taken into account.

Transworld submitted that Glencore's smaller hedging loss, by closing out its position early, should be taken into account. Glencore was required to close out its position in order to reduce exposure to the accrual of greater hedging losses against which there was no physical cargo to offset. Ignoring the hedging position would effectively give Glencore a windfall.

The court held that account was to be taken of Glencore's reduced hedging loss. Applying the approach applied in *Kaines (UK) Ltd v Osterreichische Warrenhandelsgesellschaft Austrowaren mbH,*[11] it was plain on the evidence that, having accepted Transworld's breach as bringing the contract to an end, Glencore:

> *not only did but was required to mitigate its loss by closing out its hedges. To have allowed them to run on would have been to speculate in the movement of the price of oil, which Glencore has asserted is no part of its business for present purposes.*

By closing out its hedges, Glencore established its loss. On its own evidence, hedging was an integral part of the business by which Glencore entered into the contract. Since closing out on early termination established a lower loss than would otherwise have been incurred, that had to be taken into account when determining recoverable loss.[12]

The *Glencore v Transworld* case represents an important change in how damages under international sale contracts are assessed. It emphasises the importance of hedging to damages claims under contracts for the sale of physical commodities.

Hedging losses are discussed in greater detail in sections 9–12 of this chapter.

---

10 Assuming that contracts for the delivery of Nigerian crude were typically negotiated 15–45 days prior to delivery, the question arose as to which of these days or which day to take in order to calculate market price. The solution adopted by Transworld's expert was to take an average of all the May Dated Brent swap prices. The court found that there did not seem to be any particular merit in that as opposed to another approach, and that whatever range was adopted was essentially arbitrary.

11 [1993] 2 Lloyd's Rep 1 at 10, per Bingham LJ (Court of Appeal, England & Wales).

12 Glencore was awarded damages of US$8,665,496.

7. Late delivery

Under English law, damages for late delivery of goods under a contract of sale will be assessed on the basis of the difference between the market value of the goods on the last date when they should have been delivered and market value as at the date when they were actually delivered.

Consequential losses incurred by a buyer in the case of late delivery are likely to be much the same as in the case of non-delivery. The normal measure of damages is, however, different and more analogous to the normal measure of damages for breach of warranty.

Thus, the normal measure of damages, as stated in section 53(3) of the 1979 Act, is *prima facie* the difference between the value of the goods at the time of delivery to the buyer and the value they would have had if they had been delivered on time. No rule is, however, prescribed by the 1979 Act for the measure of damages where goods sold are delivered late and ultimately accepted by the buyer after the delay.

In *Choil Trading SA v Sahara Energy Resources Ltd*,[13] the *prima facie* rule under section 53(3) of the 1979 Act was held to have been displaced. In that case, delivery to the buyer of a cargo of contaminated naphtha took place at the time of its shipment. It was not, however, until tests were carried out on board the vessel some weeks later that the contamination was discovered. As a result, the buyer's sub-buyer rejected the cargo.

The Commercial Court highlighted that there is authority that, in appropriate circumstances, a different date may be taken and that, in an FOB sale, the value may be taken at a place other than the place of shipment. The basic measure is "the estimated loss directly and naturally resulting, in the ordinary course of events, from the breach of warranty". The court held that in "a case such as this that loss (if there is a loss) is the value of the goods, had they been sound, less whatever the buyer can sell them for".[14]

8. Determination of market value – a commercial approach

The judgment in *Galaxy Energy International Ltd v Murco Petroleum Ltd (MV 'Seacrown')*[15], which involved the late delivery of 35,000 tonnes of low sulphur fuel oil sold on an FOB basis, provides some useful further guidance on the assessment of damages in international sales contract disputes.

The central question in the *Galaxy v Murco* case was how to determine the market value of the goods. As is frequently the case in the oil trade, time is critical. Sale prices are often based on a market index price at a future date linked to delivery. Murco (the oil refiner seller) argued that the market value should be determined by using the mean price across a spread of days; this was the method used by oil traders when pricing deals, so the assessment of damages should be computed using a similar method.

Galaxy (the oil trader buyer) quantified its loss conventionally, claiming the

---

13 [2010] EWHC 374 (Comm).

14 See also *Bominflot Bunkergesellschaft für Mineralöle mbH Co KG v Petroplus Marketing AG* [2012] EWHC 3009 (Comm).

15 [2013] EWHC 3720 (Comm).

difference between the market price on the last contractual date for delivery and the market price on the actual date of delivery. It argued that it was possible and appropriate to use the Platts price on the relevant date of breach to assess the market value, since this was the market price index referred to in the contract.

The parties' experts agreed that when trading companies reconcile their books ('mark to market') they use Platts[16] values for that particular day. The main disagreement between the parties' experts was whether, in order to determine the market price on a given day, one should use:

- the Platts quotation for that particular day; or
- some combination of prior and/or future quotations.

The court determined that "Platt's *[sic]* is not however a market or exchange on which a sale or purchase can be made in trading hours. When looking at market price, unlike a commodities exchange, Platt's is not itself the market", but that it was clear from the evidence that "trades in the market on the relevant day would be much more likely to be priced on a spread of Platt's than on the quoted figure for that day".

Accordingly, the court concluded that the correct method for assessing the market value was on the basis of Platts prices averaged over a number of days. Rather than taking an overly legalistic or artificial approach, the court was more inclined to look at the commercial reality of how the fuel oil should be priced.

9. Hedging losses

As discussed in the 'Hedging and derivatives' chapter by Catherine Jago, hedging is a common (but not universal) feature in the trading of various commodities, including crude oil and refined products. Hedging has been used for hundreds of years to protect traders and producers against adverse future commodity prices or even foreign exchange volatility in the market.

Trading companies habitually hedge in order not to be caught with open positions in a volatile market. It is not uncommon for trade finance banks to require trading companies to hedge as a condition to their being extended finance. It is therefore not surprising that claims for hedging losses have become an increasingly common feature in international sales contract disputes.

10. Establishing a causal link

In *Addax Ltd v Arcadia Petroleum Ltd*,[17] buyer Arcadia bought crude oil from seller Addax on an FOB basis. Pricing was based on the issuance date of the bill of lading ('BL'). The contract provided that: "In no event shall seller or buyer be liable for indirect or consequential damages".

In turn, Addax purchased crude oil from its supplier on a deferred pricing basis (14 days after the date of the BL). It then proceeded to hedge against this deferred

---

16 Platts is a provider of energy and metals information and a source of benchmark price assessments in the physical energy markets. Platts assessments are frequently used in crude and oil product trades, as well as and over-the-counter derivatives.

17 [2000] 1 Lloyd's Rep 493 (Comm).

pricing by purchasing four instruments known as 'contracts for differences' ('CFDs'). Arcadia failed to take delivery in time, following which Addax entered into futures contracts to hedge the 'outright risk' of having to pay its supplier more for the crude oil than it would receive from Arcadia.

Addax claimed the lost profit it would have made had Arcadia taken delivery in time, as well as the losses suffered as a result of entering into the CFDs on the basis of the delivery date in its contract with Arcadia and then entering into the futures contracts to protect its position upon loading. The claim was characterised as the difference between the value of the crude oil at the contractual delivery date and the value on the date it was actually delivered.

Arcadia argued that it was not liable as Addax's loss was caused by the CFDs and not the late delivery. Arcadia claimed that the loss was too remote as it was not aware of Addax's risk approach or the terms upon which it acquired the crude oil. Arcadia also argued that the losses were expressly excluded by contract because they were indirect or consequential losses.

The court agreed with Addax on its approach to the calculation of damages, which applied regardless of whether Addax made a profit or loss with its supplier. It also found that:

- the parties were "sophisticated oil traders";
- Arcadia "would have known" that Addax contracted with a third-party supplier; and
- given the "well-established trading pattern" for crude oil, it was "wholly foreseeable" that Addax would cover its position with "one of a multitude of hedging transactions available", even though "the contract instrument used may well vary from trade to trade (or possibly trader to trader)".

The court also found that even if it were necessary to take into account Addax's position with its supplier, the direct loss to Addax was "that which represents their net position [with the supplier]". In this regard, there was:

> *no sensible or commercial reason why the court should not take into account the costs of the hedging instruments ... [as] the costs of the hedging devices are an integral part of the calculation of the net position, and if the net position is a directly relevant loss, so must the hedging costs be so regarded. To extract the costs of the hedging devices is wrong in principle and has no commercial merit ...*

In the *Choil Trading* case,[18] Choil's sub-buyer rejected the naphtha on the basis that it was contaminated and Choil eventually mitigated its loss by concluding a replacement sale. Choil incurred certain losses under various hedging positions taken after its sub-buyer rejected the cargo. The court confirmed the widespread use of hedging in the trading context in the following terms:

> *Trading companies such as Choil habitually hedge in order not to be caught with open positions in a volatile market. Choil was required to do so by its trade finance bank. ... Sahara was well aware of the likelihood of Choil hedging, and the reasonableness of it doing so. It would have done the same.*

18 Note 13 above.

In both *Addax* and *Choil Trading*, there was a direct causal link between the underlying physical transaction and the hedging that was in fact carried out. Albeit not in the context of hedging, a different position was reached in *Glencore Energy UK Ltd v Cirrus Oil Services Ltd.*[19] This case concerned a claim by the seller (Glencore) against the buyer (Cirrus) for damages for non-acceptance of Ebok crude oil under sections 50(2) and (3) of the 1979 Act. After Cirrus failed to lift the oil in question, Glencore was able to terminate its contract with its physical supplier without any liability and avoided taking delivery. Glencore sought the difference between the contract price and the market price at the time when the oil ought to have been delivered.

The court had to determine the price that would be paid by a "willing buyer to a willing seller" for the oil for delivery in Ghana during the contractual delivery period. On the issue of market value, the court observed that:

> *... it is accepted that the best evidence of market value is constituted by arm's length deals actually made in the market. The last transaction effected effectively sets the benchmark.*

The assessment of the market value was, however, made more difficult, due to the lack of publicly available data for Ebok crude oil transactions.

Cirrus sought to avoid liability on the basis that the contract price/market price differential under section 50 of the 1979 Act was in reality a measure of the seller's lost profits, and that Glencore had suffered no lost profits in this case.

The court rejected this argument, holding that the price differential:

> *constitutes both a ceiling and a floor to the loss claim on the assumption that the seller had gone out into the market and sold at the date of breach. Movement in the market thereafter is then excluded from the calculation on the basis that any change in the figures affected thereby is the result of the seller's own decision to play the market.*

11. Remoteness

It is apparent from the reasoning in the cases referred to above that, as between commodity traders, it is not likely that a hedging claim would be considered, as a matter of principle, as being remote or as indirect or consequential in nature.

The question then arises as to whether hedging claims are likewise foreseeable, and hence can be claimed, where the claim is made against a carrier or shipowner. The most recent authorities on this point have not shown a consistent approach.

In *Trafigura Beheer BV v Mediterranean Shipping Company SA ('MSC Amsterdam')*,[20] Trafigura was the owner of a cargo of copper cathodes which was discharged at Shanghai. An issue arose as to the issuance of delivery orders to a fraudster, resulting in litigation in Shanghai over the question of ownership.

Trafigura eventually sued the carrier MSC under the bill of lading and claimed, among other things, the losses it incurred to keep an open position with its purchasers in relation to the copper. The Court of Appeal affirmed the finding of the judge below that the hedging losses were not foreseeable and thus not recoverable.

---

19 [2014] EWHC 87 (Comm). See also the Singaporean case of *Prestige Marine Services Pte Ltd v Marubeni International Petroleum (S) Pte Ltd* [2011] SGHC 270.

20 [2007] 2 Lloyd's Rep 62.

By contrast, a claim for hedging costs succeeded in the sale and leaseback case of *Parbulk AS v Kristen Marine SA and Anor.*[21] In that case, the court took into consideration the express provisions of the charterparty contract and Kristen's knowledge of the financing arrangements, and held that the swap costs were "breakage costs" as well as "proven expenses", in the following terms:

...the Claimant's entry into the hedging arrangements was indeed reasonably foreseeable, and hence not too remote. Compliance with the hedging obligations also cannot, in my judgment, be characterised even arguably as unreasonable, and I am unpersuaded that it matters how such requirement came about...

12. Hedging losses can be excluded by contract

It appears that hedging losses can be excluded by way of appropriately worded exclusion clauses, but such clauses must be carefully worded.

In *Choil Trading*,[22] the contract included the following provisions:

... Neither party shall be liable for any consequential, indirect or special losses or special damages of any kind arising out of or in any way connected with the performance of or failure to perform the agreement.

Neither the seller nor the buyer shall in no [sic] circumstances be liable for; –more than the difference between the contract price and the market price; –any loss of profit; –cost of overheads thrown away; or –loss resulting from shut-down of seller's plant...

The court held that the second part of the exclusion clause was applicable to cases of non-delivery or non-acceptance, but not to defective delivery:

More problematic is the ambit of the second paragraph of the clause. Like all exemption clauses it must be construed with some strictness... So construed I cannot regard it as applicable to damages for delivery of defective product... The difference between contract and market price is readily applicable in the case of non-delivery or non-acceptance, in whole or in part. It seems to me inapplicable to a claim for defective delivery where the classic measure is the difference between the market price at the date and place of delivery less the value of the goods at that date and place in their damaged state. To this comparison the contract price is not relevant... Such a conclusion derives support from the reference to "–costs of overheads thrown away; or –loss resulting from shutdown of seller's plant" which seems to refer to damages for non-acceptance.

Some widely used standard terms and conditions in the oil trade contain clauses which purport to exclude hedging losses. For example, the 2007 BP General Terms and Conditions for CFR Sales provide as follows at clause 32.1:

... Except as specifically provided ... in no event, including the negligent act or omission on its part, shall either party be liable to the other ... in contract, tort, breach of statutory duty or otherwise, in respect of any indirect or consequential losses or expenses including (without limitation) if and to the extent that they might otherwise not constitute indirect or consequential losses or expenses, loss of anticipated profits, ..., hedging or other derivative losses, goodwill, use, market reputation, business receipts or contracts or commercial opportunities, whether or not foreseeable.

---

21 [2010] EWHC 87 (Comm).
22 Note 13 above.

This clause featured in *Glencore v Cirrus*,[23] in which Cirrus unsuccessfully argued that Glencore's claim was for lost profits that were expressly excluded under clause 32.1 whether or not they were seen as "indirect or consequential losses or expenses".

Given the prevalence of hedging strategies in the physical market, and continued market volatility, it seems inevitable that losses relating to the corresponding financial trades will become an increasingly common feature in international sales contract disputes.

In addition to considering the viability of incorporating appropriately worded exclusion and limitation clauses, parties should also be mindful that they have enough evidence reasonably to support any claim for hedging losses. This may include:

- documentation attributing the hedges to specific physical trades;
- recording any steps taken to close or unwind the hedges; and
- accounting for the costs and losses in a clear manner.

Detailing the functionality of such derivative contracts or disclosing the same may, however, be problematic, particularly if issues involving confidentiality and proprietary information arise.

Consider also the not uncommon situation where a trader does not put on specific hedges but 'hedges' by balancing its sale and purchase obligations across its entire book. The identification of the exact hedging strategy and tools adopted by a trader for a specific contract can often be an extremely difficult, time consuming and costly exercise. This may require the trader to review historical systems and records, as well as daily reports over the relevant period from any risk management monitoring systems. The extremely difficult task will then be to try to extract from the entire hedging book the hedges specifically linked to the market risk associated with the contract in question, and then to show how the claimed losses can be properly attributed.

It should be noted that, in the 2015 BP GTCs (section 66) "hedging or other derivative losses" are no longer expressly excluded. As the cases considered above demonstrate, hedging losses are now, in principle, recoverable as loss, subject to the rules regarding causation, remoteness and mitigation.

13. Exclusion clauses and limiting liability

As discussed in section 12 above, it is not uncommon for commercial parties to include in their international sales contracts clauses which exclude liability for any indirect or consequential loss. The parties equally have the right to agree and to state in their contract both the type and amount of damages payable to the innocent party following the other party's breach of contract.

International sales contracts commonly include provisions which specify a liquidated or fixed amount that is payable in certain circumstances: for example, an allowance payable for a minor deficiency in quality specification.

Clauses that seek to exclude or limit a party's liability are also common in the

---

23 Note 19 above.

offshore oil and gas sector. *Transocean Drilling UK Ltd v Providence Resources PLC ('The Arctic III')*[24] involved a dispute in respect of the financial consequences of delays which occurred in the drilling of an appraisal well on Providence's Barryroe oilfield using the semi-submersible drilling rig, the 'Arctic III'. In this case, the court examined the exclusion of consequential loss and held that, in theory, a clause of this nature could operate to preclude any right of setoff, since the right cannot arise where there is no recoverable claim to setoff: "a damages claim which is excluded by contract is no claim at all".

One element of the parties' dispute focused on the meaning of "loss of use" in the context of the various other heads listed in the exclusion clause. The court construed the clause *contra proferentem* against Transocean, notwithstanding that it contained mutual exclusions and indemnities. The court's holding was made on the basis that any party relying on an exclusion clause must establish a clear intention to deprive the other party of a remedy to which it would otherwise be entitled:

> *... [O]ne starts with the presumption that neither party intends to abandon any remedies ... arising by operation of law, and clear words must be used in order to rebut that presumption.*[25]

This narrow construction was supported by the fact that other types of loss were specifically identified and therefore distinct from "loss of use".

In *Transocean Offshore International Ventures Ltd v Burgundy Global Exploration Corp*[26] [2013] 3 SLR 1017, which involved an offshore drilling contract, the Singapore High Court reiterated that the interpretation of exclusion clauses is to be done strictly, such that any ambiguity will be construed against the party seeking to rely on the clause. In construing the phrase "any indirect or consequential loss howsoever caused or arising", the court held that this only applied to such losses that fell within the second limb in *Hadley v Baxendale*.

These cases show how the courts are increasingly minded to adopt a strict approach to the interpretation of exclusion clauses by reference to business sense.

The judgment in *Glencore v Cirrus* underlines the importance of careful and unequivocal drafting of exclusion clauses and the need to specify expressly the types of liability to be excluded. Commercial parties engaged in the international sale of goods should review the exclusion clauses in their standard terms to ensure that they clearly cover all the potential losses that they are intended to cover. In particular, it is crucial for buyers of goods to check that their exclusion clauses deal separately with specific heads of loss (for example, loss of a bargain), as well as loss of profits, whether direct or consequential.

It should be noted that issues often arise in relation to 'string contracts', under which parties assume that their contracts are back-to-back, including in respect of the ability to claim or recover damages and limit liability. It is, however, often the

24 [2014] EWHC 4260 (Comm).

25 See also *Kudos Catering (UK) Ltd v Manchester Central Convention Complex Ltd* [2013] EWCA Civ 38, in which the Court of Appeal ruled that a consequential loss exclusion clause did not exclude liabilities for repudiatory breach, and should be read with the indemnities with which it was placed, such that it did not apply more generally to the contract as a whole.

26 [2013] 3 SLR 1017.

case that such chain contracts contain differently worded clauses and/or different governing law clauses. In such circumstances, there can be an increased risk of inconsistency between sale contract and sub-sale contract, specifically in relation to how contractual provisions may be interpreted and the extent to which damages are recoverable.

Joint selling and competition law

Jasvinder Nakhwal
Jonathan Tickner
Peters & Peters Solicitors LLP
Philip Woolfe
Monckton Chambers

1. Introduction

Competition law is important for any business of size, and the oil and gas sectors are no exception. Indeed, the present day landscape of multinational oil and gas operators was shaped by the 1911 break-up, under US anti-trust law, of Standard Oil.[1] Further, certain features of the sector, in particular the level of investment required for the development of new oil and gas fields and the fact that major players operate at several levels of the value chain, mean that the oil and gas sector is marked by a higher degree of cooperation between competitors than is common in many other industries.

This chapter outlines behavioural competition law, ie, the law applicable to the behaviour of undertakings, such as collusion and the abuse of market power, as it applies to the oil and gas sector in the United Kingdom and the European Union, with a particular focus on the topic of joint selling. It does not, however, cover structural competition law[2] or sector-specific regulation.[3] It also covers key aspects of procedure relating to regulatory anti-trust investigations, criminal prosecutions and private litigation.

2. Fundamental concepts

2.1 Market definition

'Market definition' identifies and defines the boundaries of competition between firms,[4] and is central to applying competition law. Unilateral conduct is only capable of infringing UK and EC competition law where an undertaking holds a 'dominant position' in a defined market. Further, market definition aids in assessing the competitive effects of agreements. The market must be defined both in its product dimension (ie, which goods or services compete) and in its geographic dimension (ie, in which territory do those goods or services compete).[5] A robust market definition

1 Successor companies to Standard Oil include ExxonMobil, Chevron and Amoco.
2 That is, merger control and instruments such as market investigations.
3 Such regulation may nonetheless be informed by competition policy, such as the EU Third Internal Energy Market Package which regulates, among other things, access to gas distribution networks and storage facilities in the interests of promoting competition and market integration.
4 See Commission Notice on the definition of the relevant market for the purposes of Community competition law, OJ 1997 C 372/03 at paragraph 2.
5 *Ibid* at paragraphs 25–31.

process requires technical economic input, but typically involves taking a candidate good or service and considering how customers and other suppliers would respond if a hypothetical monopolist of that product increased prices by 5–10% above the competitive price. If customers would switch away to other goods or services so that the price rise could not profitably be sustained, then those products are competitive constraints and must be included in the relevant market.[6] By way of example, the EU Commission has defined[7] various markets in the oil and gas sector including:

- national markets in central and eastern European countries for the wholesale supply of gas;[8]
- the market for the supply of gas to large customers in Belgium;[9]
- a market for premium and standard grade motor spirit in the Netherlands;[10]
- markets in Bulgaria for capacity in gas import pipelines, transmission pipelines and gas storage facilities;[11]
- a market for directional drilling services which is larger than national and possibly European Economic Area (EEA)-wide or worldwide in scope.[12]

In the application of anti-trust rules, although the gas markets have predominantly been defined to be national in geographic scope, the Commission is particularly concerned about agreements that tend to reinforce and perpetuate the separation of national markets,[13] in particular by overt market sharing,[14] restrictions on the resale of exported gas[15] or the restriction or reservation of gas import capacity.[16]

It is also important to consider whether an agreement is between undertakings active at the same level of production or distribution (a 'horizontal' agreement) or between undertakings active at different levels (a 'vertical' agreement).[17] Vertical restraints are generally considered to be less likely to be harmful than horizontal

6 It is also necessary to consider the possibility of swift supply-side switching.

7 Some of the market definitions that follow were provisional, in the sense that they formed the basis of statements of objection rather than concluded decisions.

8 See Case 39816, *Upstream gas supplies in Central and Eastern Europe*, Commission press release of April 22 2015, IP/15/4828.

9 COMP/B-1/37966, *Distrigaz*. The Commission noted that this market was possibly subdivided into separate markets for different types of customer such as industrial customers, electricity producers and resellers, but such subdivision was not material for the purposes of that case.

10 Decision 77/327/EEC, *ABG Oil Companies operating in the Netherlands*, OJ 1977 L 117 p1.

11 *BEH Gas*, Commission press release of March 23 2015, IP/15/4651. The Commission has also sought to promote the development of secondary markets for capacity rights: see COMP/36246 *Marathon/Ruhrgas/GDF et alia*, press release of November 23 2001, IP/01/1641.

12 Case M.5839, *Schlumberger/Smith International* (July 26 2010).

13 A long-standing general aim of European competition and energy policy is to promote the integration of national markets into wider European markets. That aim lies behind the Third Internal Energy Market Package. So far as it concerns gas, the Package comprises (1) Directive 2009/73/EC concerning common rules for the internal market in natural gas and repealing Directive 2003/55/EC, OJ 2009 L 211/94, and (2) Regulation (EC) No 715/2009 on conditions for access to the natural gas transmission networks and repealing Regulation (EC) No 1775/2005, OJ 2009 L 211/36. Other parts of the Package apply to electricity markets. It does not extend to oil.

14 COMP/39401, *E.ON/GDF*.

15 See COMP/38662, *GDF – décision GDF/ENEL*.

16 COMP/39316, *Gaz de France*.

17 See the Commission's *Guidelines on the applicability of Article 101 of the Treaty on the Functioning of the European Union to horizontal co-operation agreements*, OJ 2011/C 11/01 at paragraph 12 and its *Guidelines on Vertical Restraints*, OJ 2010/C 130/01, at paragraph 24.

restraints and to provide substantial scope for efficiencies.[18] Vertical agreements may, however, nonetheless restrict competition, eg, an agreement at the wholesale level which contains a clause aimed at preventing competition on downstream markets by restricting the resale of gas to a particular national territory.[19]

2.2 Territorial application of EU and UK competition law

The Treaty on the Functioning of the European Union (TFEU), which contains the primary provisions of EU competition law, applies within the territories of the EU member states,[20] and parallel provisions apply in respect of Iceland, Liechtenstein and Norway.[21] Articles 101 and 102 of the TFEU will only apply, however, where the agreement (or other cooperation) or the alleged abuse "may affect trade between Member States". Effect on trade may be established where it is "possible to foresee with a sufficient degree of probability on the basis of a set of objective factors of law or fact that it may have an influence, direct or indirect, actual or potential, on the pattern of trade between Member States, such as might prejudice the aim of a single market in all the Member States".[22] It is not necessary that inter-state trade is negatively affected or even actually affected; it is enough that there is potentially an appreciable effect on the pattern of trade.[23] In consequence of this approach to territorial application, it is not necessary that the agreement is actually formed, or the other conduct in question actually takes place, within the territory of the EU.[24] The Commission has published guidelines on its application of the effect on trade concept.[25]

Where member state competition authorities and courts apply national competition law, they must also consider the application of articles 101 and 102 of the TFEU if there is an effect on inter-state trade.[26] An agreement or an abuse of a dominant position may be prohibited under EU law even if it is permissible under national law. Where, however, EU law applies, an agreement that is permitted under article 101 of the TFEU may not be prohibited by national law.[27]

---

18 *Guidelines on Vertical Restraints*, paragraph 6. A specific block exemption applies in respect of certain vertical agreements, where both supplier and customer occupy less than 30% of the relevant market, subject to certain exceptions and exclusions, in particular for agreements containing 'hard core' restraints, such as resale price maintenance.

19 COMP 38662, *GDF – décision GDF/ENEL*.

20 See article 52 of the Treaty on European Union 1992 and article 355 of the TFEU.

21 The European Economic Area Agreement (EEA Agreement) applies to these states and EU Member States. Where the conduct in question is likely to affect trade between EEA member states, articles 53 and 54 of the EEA Agreement will apply instead of or in addition to articles 101 and 102 of the TFEU. In substance, articles 53 and 54 of the EEA Agreement are identical to articles 101 and 102 of the TFEU respectively. They are also enforced by the Commission. This chapter therefore addresses the EU provisions.

22 See, for example, Joined Cases C-125/07P, C-133/07 P, C-135/07P and C-137/07P, *Erste Group Bank AG and others v Commission* [2009] ECR I-8681, [2010] 5 CMLR 443 at paragraph 36.

23 See *ibid* at paragraph 36; Case 19/77, *Miller International Schallplatten GmbH v Commission* [1978] ECR 131 at paragraph 15.

24 The Commission has repeatedly stated that conduct is subject to EU law if it produces effects in the EU: see, for example, *XIth Report on Competition Policy* (1981), points 34–42. The European Court of Justice has not formally approved this 'effects doctrine', which is controversial under public international law, but has held that an agreement which is implemented in the EU is caught by article 101 of the TFEU even where the undertakings in question are based outside the EU: see Cases 89/95 etc, *Åhlström Osakeyhtiö v Commission* [1998] ECR 5193, at paragraphs 16–18.

25 *Effect on Trade Guidelines*, OJ 2004 C101/81.

26 Article 3(1) of Regulation 1/2003, OJ 2003 L1/1.

27 *Ibid*, article 3(2). Unilateral conduct which is permissible under article 102 of the TFEU may, however, nonetheless be prohibited by national law.

The UK has adopted an approach to determining the territorial application of its competition law similar to that which has been adopted by the EU. The Chapter I and Chapter II prohibitions will apply where the agreement or putative abuse "may affect trade within the United Kingdom".[28]

3. Anti-competitive agreements in UK and EU law

Article 101 of the TFEU and its analogue in UK law, Chapter I of the Competition Act 1998, aim at prohibiting cooperation between undertakings which has as its object or effect the prevention, restriction or distortion of competition, unless that co-operation is shown to provide countervailing efficiencies. The EU and UK provisions comprise both a basic prohibition and a provision for exemption: agreements which fall within the basic prohibition and which are not exempted are prohibited. The language of the UK provisions is modelled on, and is required to be interpreted so far as possible to be consistent with,[29] the EU provisions.

The prohibitions catch not only formal agreements but also informal cooperation, including "all agreements between undertakings, decisions by associations of undertakings and concerted practices."[30] The concept of a concerted practice is very broad and has been held to cover any "form of coordination between undertakings which, without having been taken to a stage where an agreement properly so called has been concluded, knowingly substitutes for the risks of competition practical cooperation between them".[31] Thus it can include not only, for example, an agreement or understanding on prices, but also the exchange of competitively sensitive information where the likely effect of that exchange is to reduce competition.[32] It includes indirect 'hub-and-spoke' arrangements, eg, where information is shared between competitors via a customer or supplier.[33]

Only agreements between undertakings are caught by the prohibition provisions. An 'undertaking' has been defined as any entity engaged in offering goods and services in a given market, regardless of its legal status and the way in which it is financed,[34] including state-owned bodies, such as Statoil.[35] It does not, however, include government bodies that are not active in a market.

3.1 The basic prohibition

Under both EU and UK law, two types of agreements fall within the basic prohibition: restrictions by object and restrictions by effect. Restrictions by object are agreements which 'by their nature' restrict competition, such that the relevant competition authority or court is not required to examine whether the agreement in

---

28 See sections 2(1)(a) and 18(1) of the Competition Act 1998.

29 See section 60 of the Competition Act 1998.

30 Article 101(1) of the TFEU.

31 Case C-49/92P, *Commission v Anic Partecipazioni SpA* [2001] 4 CMLR 17, at paragraph 115.

32 See *United Kingdom Agricultural Tractor Registration Exchange*, OJ 1992 L68/19, on appeal Case T-34/92, *Fiatagri and New Holland Ford* [1994] ECR II-905.

33 See, for example, the Office of Fair Trading decisions in *Toys and Games* (February 19 2003) and *Replica Football Kit* (August 1 2003).

34 Case C-205/03, *Federación Española de Empresas de Tecnología Sanitaria (FENIN) v Commission* [2006] ECR I-6295.

35 See *FENIN* and the *GFU* case, discussed below.

fact had the effect of restricting competition.[36] If it is not plain that the object of the agreement is to restrict competition, it then becomes necessary to consider its effects, taking into account the economic context in which the undertakings operate, the products or services covered by the agreement, the structure of the market concerned and the actual conditions in which it functions.[37] In analysing the effect on competition, it is necessary to consider the competition that would occur in the absence of the agreement in dispute. In order for article 101(1) of the TFEU or section 2 of the Competition Act 1998 to apply, it has to be shown that the effect on competition is appreciable.[38]

3.2 Exemption

If an agreement is found to infringe the basic prohibition, it is necessary to go on to consider whether the agreement is exempted, either individually[39] or under an EU 'block exemption' regulation. The provisions governing individual exemption are essentially identical in EU and UK law.[40] In order to benefit from individual exemption it is necessary to show that four conditions are met, namely that the agreement:[41]

- contributes to improving the production or distribution of goods or to promoting technical or economic progress;
- allows consumers a fair share of the resulting benefits;
- does not impose restrictions which are not indispensable to the attainment of these objectives; and,
- does not afford such undertakings the possibility of eliminating competition in respect of a substantial part of the products concerned.

The EU Commission has published regulations exempting certain categories of agreement. These exemptions are subject to detailed conditions, eg, in respect of market share or requirements that the agreement does not contain terms of a specified type. Where there is a risk that an agreement may infringe the basic prohibition but may fall within the scope of a block exemption, it is therefore necessary to make a careful analysis of the terms of the agreement and the relevant block exemption. These block exemptions are given parallel application in UK law.[42]

---

36 *Institute of Independent Insurance Brokers v Director General of Fair Trading* [2001] CAT 3, [2003] ECC 25 (*IIB*), at paragraph 169, relying on Case 56/65, *Société Technique Minière* [1966] ECR 235 at 249. See also Cases 56 and 58/64, *Ets Consten SaRLand Grundig-Verkaufs-GmbH v Commission* [1966] ECR 299, at 342; Cases T- 25/95 etc, *Cimenteries CBR v Commission* [2000] ECR II-491, paragraph 1531.

37 See *IIB*, at paragraph 170 relying on *Société Technique Minière* at 249 to 250; Case C-399/93 *Oude Luttikhuis v Verenigde Coöperatieve Melkindustrie Coberco BA* [1995] ECR I-4515, paragraph 10 and Cases 180 to 184/98 *Pavlov v Stichting Pensioenfonds Medische Specialisten* [2000] ECR I-6451, paragraph 91

38 *IIB* at paragraph 171 relying on Case 5/69, *Völk v SPRL Ets J Vervaeke* [1969] ECR 295, paragraph 7; Case 22/71, *Béguelin Import Co v SA GL Import Export* [1971] ECR 949, paragraph 16.

39 Agreements must be self-assessed. By contrast with the situation prior to May 2004, it is no longer possible to apply to the EU Commission or the UK competition law authorities for a decision that an agreement is exempt (or does not infringe the basic prohibition).

40 See article 101(3) of the TFEU and section 9 of the Competition Act 1998.

41 Article 101(3) of the TFEU; section 9(1) of the Competition Act 1998. The EU Commission has published extensive guidance on the application of these criteria: see *Guidelines on the application of Article 101(3) TFEU (formerly Article 81(3) TEC)*, OJ 2004 C101/97.

42 See section 10 of the Competition Act 1998.

The block exemption regulations which are potentially of relevance to the oil and gas sector cover:[43]

- agreements for cooperation in research and development;[44]
- 'specialisation agreements' (ie, agreements under which one party agrees fully or partly to cease production of certain products and to purchase them instead from the other party);[45] and
- vertical agreements.[46]

3.3 Pure and hybrid joint selling arrangements

'Pure' joint selling agreements are likely to infringe the basis prohibition in article 101(1) of the TFEU or section 9 of the Competition Act 1998 and are unlikely to meet the conditions for individual or block exemption. This is because, if undertakings could sell the product in question individually, but instead agree to sell it jointly, the agreement would eliminate the competition that would otherwise occur between the parties.[47] For example, the Commission has pursued cases against national selling arrangements for the joint selling of gas in both Norway and Denmark.

From 1989 to 2002, Norwegian gas was sold to the EU via a Gas Negotiation Committee made up of the Norwegian producers (the GFU), which fixed prices, volumes and other trading conditions. The European Commission issued a statement of objections in respect both of that joint selling and of the long term contracts entered into by committee members with third parties, which the Commission contended perpetuated the adverse effects of the joint selling. Proceedings were discontinued after Statoil and Norsk Hydro gave binding commitments to cease joint marketing and sales activity and to reserve some gas volumes for new customers.[48]

The Commission also investigated the joint marketing of gas originating in Denmark by Shell, ChevronTexaco and AP Møller through the Danish Underground Consortium (DUC). The parties initially claimed that the joint selling was exempt as a 'specialisation' agreement under the block exemption. The arrangement was, however, terminated by agreement with the Commission following the GFU case, as were restrictive terms in agreements between DUC members and the incumbent Danish national gas supplier (DONG) which provided for information sharing and discouraged DUC members from entering the market for gas supply in Denmark.[49]

---

43 A detailed description of the terms of the block exemptions is beyond the scope of this work, but their specific relevance to joint selling arrangements is considered below. Sector-specific block exemption regulations have been adopted in other areas.

44 Commission Regulation (EU) No 1217/2010 of December 14 2010 on the application of article 101(3) of the Treaty on the Functioning of the European Union to certain categories of research and development agreements, OJ 2010 L335/36.

45 Commission Regulation (EU) No 1218/2010 of December 14 2010 on the application of article 101(3) of the Treaty on the Functioning of the European Union to certain categories of specialisation agreements, OJ 2010 L335/43.

46 Commission Regulation (EU) No 330/2010 of April 20 2010 on the application of Article 101(3) of the Treaty on the Functioning of the European Union to categories of vertical agreements and concerted practices, OJ 2010 L102/1.

47 See the Commission's decision in *Floral*, OJ 1980 L39/51, [1980] 2 CMLR 285; *Guidelines on the applicability of Article 101 of the Treaty on the Functioning of the European Union to horizontal co-operation agreements*, OJ 2011 C11/1 (*'Horizontal Co-operation Guidelines'*) at paragraphs 234–235.

48 See COMP 36072, *GFU*, Press Release IP/02/1084 (July 17 2002).

49 See COMP 38187, *PO/DUC-DONG*, Press Release IP/03/566 (April 24 2003).

Nonetheless, even pure joint selling arrangements may in principle be permissible. If the undertakings in question could not enter, or continue, in the market separately, then the agreement would not eliminate any competition that would occur in the absence of the agreement.[50] Similarly, a joint selling agreement may be held not to infringe the basic prohibition where, even though the undertakings could sell their rights individually, joint selling was necessary to promote the entry of a new purchaser into downstream markets.[51] Further, even if competition is restricted, the arrangement may be exempt if it leads to genuine efficiencies, such as cost savings.[52] Merely increasing the negotiating power *vis-à-vis* incumbent energy companies is, however, unlikely to be considered a genuine efficiency so as to justify joint selling.[53]

Parties may also engage in joint selling as part of a wider arrangement involving joint research and development, joint production or specialisation. Where joint selling is necessary to incentivise the parties to enter into a production agreement, the joint setting of sales prices will not be treated as a restriction of competition by object and the agreement must be assessed as a whole for its effect on competition.[54] There is not an anti-competitive effect merely because one of the parties ceases to produce the product, provided that there are several alternative actual or potential suppliers and spare capacity.[55] The potentially anti-competitive effects of such joint production agreements include:

- the elimination of the possibility of competition between the parties (including in respect of production such as on cost minimisation or technical development, as well as on price);[56]
- the agreement may constitute collusion if the parties have market power and the market structure is conducive to co-ordination;[57] and
- the agreement may lead to anti-competitive foreclosure of third parties in a related market, in particular where the product is a key input in a downstream market and cooperation would permit the parties to raise the costs of rivals in that market.[58]

Joint selling arrangements may be block exempted where they are ancillary to a

---

50 See Case 56/65, *Société Technique Minière* [1966] ECR 235 at 250. This case concerned an exclusive distribution agreement.

51 See *Bookmakers Afternoon Greyhound Services Ltd v Amalgamated Racing Ltd* [2009] EWCA Civ 750, which involved an agreement for the joint sale of media rights.

52 See *Horizontal Co-operation Guidelines* at paragraphs 246–247. Cost savings for the undertakings in question which arise solely from the reduction in competition do not count for these purposes.

53 In COMP 37708, *Irish Corrib gas field*, the joint owners of the field applied for an exemption covering a joint marketing arrangement for the first five years of its operation. The companies argued that joint marketing would be necessary to balance the countervailing purchasing power of the incumbent Irish energy companies. Following the expression of concerns by the Commission, however, the application was withdrawn: see press release IP/01/578 (April 20 2001).

54 *Horizontal Co-operation Guidelines*, paragraph 160.

55 See *Philips/Osram*, OJ 1994 L378/37, point 17.

56 *Horizontal Co-operation Guidelines*, paragraphs 157, 164 and 187; *Olivetti/Canon*, OJ 1988 L 52/51, at point 42.

57 *Horizontal Co-operation Guidelines* at paragraphs 158 and 165. This is particularly true if the co-operation creates a commonality of costs or involves information sharing.

58 *Horizontal Co-operation Guidelines* at paragraphs 159 and 165.

specialisation or joint production agreement that falls within Regulation 1218/2010. This regulation applies both to unilateral and reciprocal specialisation agreements as well as to joint production agreements.[59] Importantly, the block exemption extends to specialisation agreements that provide for joint distribution.[60] It is a condition of application of the block exemption regulation that the combined market share of the parties does not exceed 20% in any relevant market.[61] It is important to note that market shares must be calculated taking into account connected undertakings,[62] and on the basis of market sales data or, if such data is not available, other reliable market information, and in each case on the basis of data relating to the preceding calendar year.[63] Agreements will fall outside the scope of the block exemption if they contain provisions that have certain 'hard core' restrictions as their direct or indirect object.[64] First, agreements may not have as their object the fixing of prices when selling the products to third parties. Agreements which provide for joint distribution may, however, set the prices to be charged to immediate customers.[65] Secondly, agreements may not have the object of limiting output or sales. This is, however, subject to three exceptions:

- unilateral and reciprocal specialisation agreements may include provisions on the agreed amounts of products;
- joint production agreements may set capacity and production volume; and
- agreements which provide for joint distribution may set sales targets.[66]

Finally, agreements may not have as their object the allocation of markets or customers.[67]

---

59 See articles 1.1(a) and 2.1 of Regulation 1218/2010. A unilateral specialisation agreement is an agreement between two parties, active in the same product market, by virtue of which one party (A) agrees to cease production of certain products fully or partly or to refrain from producing those products and also to purchase those products from the other party (B), who agrees to produce and supply those products to A. A reciprocal specialisation agreement involves both A and B agreeing to give up production of different goods and instead to purchase these goods from one another.

60 Article 2(2)(b) of Regulation 1218/2010. Joint distribution includes the appointment of a third-party distributor on an exclusive or non-exclusive basis, provided that the third party is not a competing undertaking: see Regulation 1218/2010, article 1.1(q). It should also be noted that the block exemption can extend to agreements containing (1) ancillary provisions for the assignment or licensing of intellectual property rights if those provisions do not constitute the primary object of the agreement, and (2) obligations for exclusive purchasing or exclusive supply: see articles 2(2) and (3)(a) of Regulation 1218/2010.

61 See article 3 of Regulation 1218/2010. For these purposes, 'relevant markets' are the relevant product and geographic markets to which any specialisation product belongs and, in addition, where the specialisation products are intermediary products that one or more of the parties fully or partly use for the production of downstream products, the relevant product and geographic markets to which the downstream products belong: see articles 1.1(i)–(k) and 3 of Regulation 1218/2010. Exemption may continue to apply for two calendar years where the market share, having initially been not more than 20%, subsequently rises above that level but does not exceed 25% (see *ibid*, article 5(d)) and for one calendar year where the market share, having initially been not more than 20%, subsequently rises above 25% (see *ibid*, article 5(e)).

62 *Ibid*, article 1.2 and 5(c). The definition of 'connected undertakings' in the Block Exemption Regulations is complex and includes both direct and indirect controlling and controlled undertakings.

63 See article 5 of Regulation 1218/2010.

64 'Object' must be judged both from the terms in isolation and in combination with other factors under the control of the parties: see *ibid*, article 4.

65 *Ibid*, article 4(a).

66 *Ibid*, article 4(b).

67 *Ibid*, article 4(c).

Similarly, the protection of the Research and Development Block Exemption Regulation[68] extends to certain aspects of joint selling where it is directly related to and necessary for the implementation of a research and development agreement and does not constitute the primary object of the contract.[69] Where a research and development agreement provides for joint distribution, the agreement may provide for sales targets and the fixing of the selling prices.[70] Where the joint exploitation of the results of the research and development includes joint production, the parties may also agree on production targets.[71] If the parties are actual or potential competitors, it is a condition of application of the block exemption regulation to such agreements that the combined market share of the parties does not exceed 25% in any relevant product or technology market.[72]

Where joint production and joint selling are implemented through the creation of a 'full-function' joint venture,[73] that joint venture will fall to be assessed as a merger under UK or EC law.[74] The treatment of such mergers falls outside the scope of this work. It may, however, be noted that, from a commercial perspective, a significant advantage of a transaction being structured as a full-function joint venture is that the transaction may be notified to the competition authorities in advance for clearance, whereas lesser forms of cooperation must be self-assessed by the parties and run the risk of later being found by the competition authorities to have been unlawful.

4. Article 102 of the TFEU

In addition to the prohibition on anti-competitive collusion described above, UK and EU competition law specifically regulate conduct by undertakings with market power. Article 102 of the TFEU and Chapter II of the Competition Act 1998 therefore prohibit the "abuse by one or more undertakings of a dominant position" within the internal market or within the UK respectively.[75] These provisions therefore cover the same broad subject matter as section 2 of the US Sherman Act, but do so using very different language and concepts.[76]

A 'dominant position' refers to "a position of economic strength enjoyed by an undertaking which enables it to prevent effective competition being maintained on

68 Regulation 1217/2010.

69 *Ibid*, article 3(2).

70 *Ibid*, articles 5(b)(ii) and 5(c).

71 Article 5(b)(i) of Regulation 1217/2010.

72 *Ibid*, article 4(2)(a). The market share threshold applies subject to a seven-year transitional period in the case of undertakings that are not actual or potential competitors at the outset of the agreement: see *ibid*, article 4(1). The threshold also applies in a different manner to paid-for research and development (see *ibid*, article 4(2)(b)), but such agreements are unlikely to involve joint selling. The provisions in respect of the calculation of market shares are essentially the same as those under Regulation 1218/2010 described above: see article 7 of Regulation 1217/2010.

73 That is, "a joint venture performing on a lasting basis all the functions of an autonomous economic entity": see Regulation 139/2004 OJ 2004 L 24/1, at article 3(4).

74 With regard to EU law, see the Merger Regulation (Reg 139/2004, OJ 2004 L 24/1). With regard to UK law, see Part 3 of the Enterprise Act 2002.

75 As under article 101 of the TFEU and Chapter I of the Competition Act 1998, an abuse is covered by EU law insofar as it has an effect on trade between member states, and by UK law insofar as it has an effect on trade within the United Kingdom: see article 102 of the TFEU and section 18(1) of the Competition Act 1998.

76 Section 2 of the Sherman Act prohibits monopolising or attempting to monopolise trade.

the relevant market by giving it the power to behave to an appreciable extent independently of its competitors, customers and ultimately of its consumers".[77] Establishing dominance requires the relevant market first to be defined as explained at section 3 above. Undertakings may be collectively dominant, and therefore subject to article 102 of the TFEU and Chapter II of the Competition Act 1998, even where they are not individually dominant. Such a position of collective dominance may arise where the undertakings concerned present themselves or act together in a particular market as a collective entity, for instance by joint selling, and where that collective entity enjoys a position of dominance.[78]

The European Court of Justice (ECJ) has described the concept of an abuse as being:[79]

> *an objective concept relating to the behaviour of an undertaking in a dominant position which is such as to influence the structure of a market where, as a result of the very presence of the undertaking in question, the degree of competition is weakened and through recourse to methods which, different from those which condition normal competition in products or services on the basis of the transactions of commercial operators, has the effect of hindering the maintenance of the degree of competition still existing in the market or the growth of that competition.*

In order to establish an abuse, it is not generally necessary to show anti-competitive intent,[80] only to demonstrate a potentially anti-competitive effect.[81] It is not necessary that the anti-competitive effect arises in the same market in which the undertaking holds its position of dominance.[82] Abuses are conventionally divided into exclusionary abuses (which protect the undertaking's position of dominance) and exploitative abuses (which leverage the dominant position to earn unfair or supra-competitive profits).

The categories of abuse are not closed and a full survey of the different types of conduct that have been held to be abusive falls outside the scope of this work. However, categories of abuse which are particularly relevant to joint selling include the following.

(a) ***Predatory pricing***

'Predatory pricing' has been defined as "the practice where a dominant company lowers its price and thereby deliberately incurs losses or foregoes profits in the short run so as to enable it to eliminate or discipline one or more rivals or to prevent entry by one or more potential rivals".[83] Its boundaries are contentious, in particular as to how it is to be distinguished from low pricing in response to competition and what cost benchmark should be used to identify predatory conduct. As a matter of

---

77 Case 27/76, *United Brands Co v Commission* [1978] ECR 207 at paragraph 65.

78 See Joined Cases C-395/96P and C-396/96P, *Compagnie Maritime Belge Transports SA* [2000] ECR I-1365 at paragraph 36.

79 Case C-62/86, *AKZO Chemie BV v Commission of the European Communities* [1991] ECR I -3359 at paragraph 69.

80 See Case C-549/10P, *Tomra Systems ASA v Commission* [2012] 4 CMLR 1093 at paragraph 21.

81 See Case C-52/09, *Konkurrentsverket v TeliaSonera Sverige AB* [2011] 4 CMLR 951 at paragraph 64.

82 *Ibid*, at paragraphs 86-88.

83 See *DG Competition discussion paper on the application of Article 82 of the Treaty to exclusionary abuses* (December 2005) at paragraph 93.

authority, however, there is a rebuttable presumption that pricing below average variable cost is predatory,[84] and pricing between average variable cost and average total cost may be considered abusive if intention to eliminate a competitor is shown.[85]

(b) ***Price discrimination and targeted price cutting***

Discrimination consists in the application of dissimilar conditions to equivalent transactions.[86] It does not include charging a different price where transactions are not equivalent, eg, where the underlying costs involved are different. The targeting of price cuts towards certain customers with the intention of eliminating a competitor has been held to constitute an abuse of a dominant position.[87] Not all price discrimination is abusive, however, in particular if:

- it would not be liable to exclude a competitor which is at least as efficient as the dominant firm; or
- the price discrimination is objectively justified by efficiency gains.[88]

(c) ***Loyalty rebates and discounts***

Loyalty rebates and discounts are not inherently problematic, but such rebates can exclude competitors from the market in some circumstances when applied by a dominant undertaking. Where there is an actual exclusionary effect, the rebates or discounts will constitute an abuse unless objectively justified.[89] Whether there is an exclusionary effect must be judged by all the circumstances, but particular reference will be made to the strength of the incentive effect of the rebate or discount.[90] Rebates or discounts that take account of cost savings driven by increasing volumes,[91] or which reflect customer-specific investments[92] may in particular be objectively justified.

(d) ***Exploitative pricing***

The pricing practices discussed above are all 'exclusionary' abuses, aimed at removing competitors, but 'exploitative' pricing may also be abusive if it is excessive in relation to the applicable costs and "either unfair in itself or when compared to competing products".[93] The vagueness of the test means, however, that such excessive pricing is difficult to identify precisely,[94] and cases are rarely brought.

---

84 See Case C-202/07P, *France Télécom SA v Commission* [2009] ECR II-2369. This judgment moved away from the earlier case law following Case C-62/86, *AKZO Chemie BV v Commission* [1991] ECR I-3359, in which the ECJ had stated that prices below average variable costs must always be considered abusive.

85 Case C-62/86 *AKZO Chemie BV v Commission* [1991] ECR I-3359, paragraphs 70–72.

86 See article 102(c) of the TFEU.

87 See Cases C-395 & 296/96P, *Compagnie Maritime Belge Transports SA v Commission* [2000] ECR I-136.5.

88 See Case C-209/10, *Post Danmark A/S v Konkurrencerådet*, [2012] 4 CMLR 23.

89 See the General Court's decision in Case T-286/09, *Intel Corp v Commission* [2014] 4 CMLR 9 at paragraph 70. Pre-existing case law of the (superior) Court of Justice, in particular, Case C-549/10P, *Tomra Systems ASA v Commission* [2012] 4 CMLR 1093 had, however, indicated that a *potential* exclusionary effect was sufficient.

90 See Case C-95/04P, *British Airways plc v Commission* [2007] ECR I-2331 at paragraph 68 and Case 322/81, *NV Nederlandsche Banden Industrie Michelin v Commission* [1983] ECR 3461 at paragraphs 81-84.

91 See *BPB Industries and British Gypsum*, OJ 1992 C321/9.

92 See *Guidance on the Commission's enforcement priorities in applying Article 82 of the EC Treaty to abusive exclusionary conduct by dominant undertakings*, OJ 2009 C45/7 at paragraph 46.

93 See Case 27/76, *United Brands Co v Commission* [1978] ECR 207, [1978] 1 CMLR 429, at paragraph 252. Unfair pricing is specifically mentioned in article 102(a) of the TFEU.

94 See *Albion Water Ltd v Water Services Regulation Authority* [2008] CAT 31 at paragraph 216.

(e) ***Margin squeeze***

Where a dominant undertaking supplies customers who resell or otherwise use the goods or services in a downstream market in competition with it, the dominant undertaking may commit an abuse if it raises the price at which it sells to those customers so that they cannot effectively compete in the downstream market, notwithstanding that they are as efficient as the dominant undertaking.[95]

(f) ***Refusal to supply***

Similarly, a refusal to supply, in particular to established customers who are downstream competitors, may be an abuse where it is discriminatory, is liable to cause those competitors to exit the market and is not objectively justified. For example, BP was found to have abused a dominant position during the 1973 oil crisis where it reduced its supplies to a purchasing cooperative of petrol and diesel which competed with it in the downstream filling station market, to a greater extent than it reduced supplies to its other customers.[96] It is, however, unlikely that a refusal to supply a new customer will be an abuse unless the good or service in question is an "essential facility" for any competition in the downstream market to occur.[97]

(g) ***Long term supply agreements***

Long term supply agreements may exclude suppliers wishing to enter the market and rigidify market shares. In certain cases, the Commission has required such agreements in the oil and gas sector to be terminated.[98]

(h) ***Tying and bundling***

'Tying' refers to situations where customers that purchase one product (the tying product) are required by contractual conditions or technical reasons also to purchase another product from the dominant undertaking (the tied product). 'Bundling' covers situations where products are sold jointly (or heavily discounted when purchased jointly, so as to encourage joint purchasing).[99] These are, however, highly contested categories of abuse, since it is particularly difficult to determine whether there are really two distinct products on offer, or one joint product.[100]

5. Consequences of breach

Enforcement of competition law in both the UK and the EU is primarily based on two key areas, namely:

---

95 See Case C-52/09, *Konkurrentsverket v TeliaSonera Sverige AB* [2011] 4 CMLR 951 at paragraphs 31 and 97.

96 *ABG oil companies operating in the Netherlands*, Decision 77/327/EEC OJ 1977 L117/1.

97 See Case C-7/97, *Oscar Bronner GmbH v Mediaprint Zeitungs- und Zeitschriftenverlag GmbH & Co* [1998] ECR I-7791, [1999] 4 CMLR 112.

98 See COMP/B-1/3796,6 *Distrigaz*.

99 See *Guidance on the Commission's enforcement priorities in applying Article 82 of the EC Treaty to abusive exclusionary conduct by dominant undertakings*, OJ 2009 C45/7 at paragraph 48.

100 See, for example, Case T-201/04 *Microsoft Corp v Commission* [2007] ECR II-3601, [2007] 5 CMLR 846, where it was disputed whether, in circumstances where there was no demand for operating systems which did not include media playing capability, it was an abuse to include the Windows Media Player with the Windows operating system. The General Court held that the question was whether there was in fact a separate market for the tied product, ie, media players, and held that there was on the basis that there were companies producing and selling such media players separately from operating systems.

- public enforcement by the established regulators; and
- private enforcement by parties to a competition law issue.

Criminal investigations may also exist in parallel with investigations into civil breaches of competition law. Since the modernisation of the enforcement of EU competition law throughout the EU pursuant to Regulation 1/2003, national courts and competition authorities have jurisdiction to apply articles 101 and 102 of the TFEU directly and are required to apply those provisions when applying national competition law.[101]

5.1 Regulatory investigations

In the UK, the Competition and Markets Authority (CMA) has general responsibility for applying both EU and UK competition law, though sectoral regulators also have concurrent powers within specific sectors of the economy,[102] including the Gas and Electricity Market Authority (GEMA).[103] Further, the European Commission is empowered to apply and enforce articles 101 and 102 of the TFEU.[104]

(a) Procedure in the EU

The Commission is responsible for fact-finding, making infringement decisions and imposing penalties under EU competition law. It may commence investigations on its own account, following receipt either of a complaint or a leniency application by an undertaking involved in suspect conduct.

In an investigation, the Commission has extensive powers of inspection to enter and seal premises, examine records and obtain copies, and interview any member of staff or representative of the party in question.[105] Following an initial investigation, the Commission may decide to proceed to an in-depth investigation, or to close the matter, on the basis of its own priorities. In December 2015, for example, it was announced that the Commission had ended a two-year investigation into Shell, BP and Statoil in relation to the suspected rigging of crude oil benchmarks, though producers Abengoa, Alcogroup and Lantmännen were subsequently included in an investigation into the manipulation of ethanol benchmarks. In the course of the former investigation, the premises of Shell, BP and Statoil were raided by EU officials looking for evidence that they had conspired to provide false information to Platts, a publisher of benchmark prices, which would then have influenced data underpinning billions of dollars' worth of trading in oil and gas.

If the Commission wishes to proceed to a formal decision after investigation, it must issue a statement of objections (SO) detailing its concerns to the party in question, which will then be given access to its case file, the right to reply and (on

101 See article 3(1) of Regulation 1/2003. This is subject to the effect on trade criterion discussed above.

102 The Competition Act 1998 (Concurrency) Regulations 2014 (SI 2014/536) and memoranda of understanding between the CMA and the sectoral regulators determine the allocation of cases between regulators.

103 GEMA is supported by the Office of Gas and Electricity Markets (OFGEM).

104 Its powers are conferred by articles 101 and 102 of the TFEU and by Regulation 1/2003. These are outlined below.

105 Article 20, Council Regulation (EC) 1/2003. It may also instruct national competition authorities in member states to undertake inspections.

request) an oral hearing. The Commission may then issue a decision finding an infringement of articles 101 or 102 of the TFEU and imposing a fine. Alternatively, it may accept commitments proposed by the party in question, which will be made legally binding by the Commission decision.

(b) *Fines*

The UK competition authorities[106] have the power to impose a fine of up to 10% of an undertaking's turnover in the previous business year.[107] The CMA has issued detailed guidance on the method of calculating the level of a fine for an infringement, taking account of:

- the seriousness of the infringement and the turnover of the undertaking in the relevant product and geographic markets in the previous business year;
- the duration of the infringement;
- aggravating and mitigating factors and the requirements of proportionality and deterrence; and
- specific mitigation.

Fines may be further reduced for reasons of leniency, settlement, or (exceptionally) when the party is unable to pay. Appeals to the CAT against regulatory findings of breach or fines must be commenced within two months of notification or publication of the relevant decision.[108]

The Commission also has power to impose fines of up to 10% of annual turnover. It sets fines by first taking a percentage of the undertaking's annual sales of the relevant product in the most recent full year of infringement, up to a maximum of 30% depending on the seriousness of the infringement. This is adjusted for the duration of the infringement, aggravating and mitigating factors, and (for cartels) an increase of 15–25% of one year's sales of the relevant product. The final fine is adjusted if necessary to prevent it from exceeding the maximum permitted fine of 10% of the undertaking's total annual turnover. Further adjustment may be made under the Commission's leniency regime or for settlement.

In the oil and gas sector, the Commission's fine of €640m against E.ON/GDF for their collusion in European gas markets is one of the highest cartel fines in EU history. This figure follows a reduction of the original €1.1bn fine by 42% following a European Court judgment which held that the Commission had overestimated the duration of the infringements.

(c) *Immunity and leniency*

In the UK, undertakings that are involved in cartel activity may apply for immunity or leniency in respect of potential civil liability. Likewise, individuals may apply for immunity from criminal prosecution for the criminal cartel offence (see below).

---

106 The CMA and several sectoral regulators, including GEMA and OFGEM.

107 Section 36 of the Competition Act 1998; Competition Act 1998 (Determination of Turnover for Penalties) (Amendment) Order 2000 (SI 2004/1259).

108 The High Court of England & Wales also has judicial review jurisdiction over regulators' decisions, so far as they are not subject to a right of appeal.

Immunity confers complete protection on a party from fines or prosecution in the case of an individual, and a successful application for leniency may substantially reduce any fines imposed on an undertaking. Applications may be made to the CMA for domestic cases and otherwise to the Commission. Applications should be made to the Commission whenever there is any possibility that it will commence an investigation, eg, because the violation involves more than one EU member state. In cases where the Commission undertakes an investigation into cartel conduct which includes the UK, the CMA will not conduct a separate civil investigation. In these circumstances, it is still possible that the CMA will carry out a criminal investigation against relevant individuals, as the Commission does not have powers to impose criminal liability against individuals for alleged cartel conduct. This remains the subject of applicable domestic legislation. Where an undertaking is granted immunity by the Commission, the CMA will ordinarily grant criminal immunity to that undertaking's past and present employees, as it would have done had it led a civil investigation and granted immunity to the undertaking in the first instance. In circumstances where an undertaking does not have immunity from civil fines, or chooses not to seek such immunity, it is still possible for an individual who is/was employed by that undertaking to obtain immunity from criminal prosecution of his/her own volition.

Several categories of immunity and leniency are available to an undertaking in the UK. They confer different benefits and are accompanied by different requirements.

Type A immunity: Type A immunity confers total protection from fines and will be granted to an undertaking if it is the first party to notify the CMA of cartel activity, the CMA does not already have enough evidence to prove the existence of cartel activity (or is unaware of the conduct), and the undertaking complies with several other requirements. These are that the undertaking must fully cooperate with the investigation, have no further involvement in the cartel, and must not at any time have attempted to coerce another undertaking to join the cartel.

When an undertaking is granted Type A immunity, all cooperating current and former employees of the undertaking are automatically granted guaranteed protection from criminal prosecution in connection with the cartel. At the initial stages of the matter, and for some time afterwards, the grant of immunity to such employees will be conditional upon a number of matters, including full and continuing cooperation. In essence, the immunity crystallises when the individual is provided with confirmation from the CMA, by way of a no-action letter.

Type B immunity: Type B immunity confers the same protections as Type A immunity, but is granted only at the discretion of the CMA. It is available to an undertaking that is the first to notify the CMA of cartel activity, but only does so after the CMA has already started an investigation and before a statement of objections has been issued. The new information must add 'significant value', the undertaking must comply with the same additional requirements as those applicable to Type A immunity,[109] and there must be no Type A applicant.

---

109 *Ibid.*

As with Type A immunity, under Type B immunity all cooperating current and former employees of the undertaking are granted guaranteed protection from criminal prosecution in connection with the cartel. However, where the CMA does not wish to confer 'blanket' immunity from criminal prosecution to employees it may, in the same circumstances as full Type B immunity, grant full corporate immunity from fines and offer individual criminal immunity only on a case-by-case basis.

Type B Leniency: Type B leniency is also available in the same circumstances as Type B immunity and, at the discretion of the CMA, may reduce the level of any fine by up to 100%. It does not automatically confer immunity from prosecution on all past and present employees, but they may be granted individual immunity on a case-by-case basis.

Type C Leniency: Type C leniency confers a reduction in the level of any fine by up to 50% and is available at the discretion of the CMA to any undertaking that is not the first to report the cartel activity, or is otherwise ineligible for a higher tier of leniency or immunity. For Type C leniency to be available to an undertaking, it must, as in the case of Type B immunity, provide information that adds significant value to the CMA's investigation. It must also comply with the same further obligations as under Type A and Type B immunity, except for the obligation that it must not have coerced any party to join the cartel.

The European Commission: The requirements for immunity or leniency and available protections discussed above in respect of the CMA are fundamentally the same for applications made to the Commission. The equivalent test for Type B immunity requires that information provided by an undertaking enables the Commission to prove the cartel infringement, and the equivalent test for Type C immunity requires evidence that represents 'significant added value' in respect of its ability to prove the infringement. In the latter case, the first undertaking to meet these conditions will be granted a 30–50% fine reduction, the second a 20–30% reduction, and further undertakings a reduction of up to 20%.

Criminal immunity: In addition to immunity being available to all past and present employees of undertakings that receive immunity, individuals may themselves apply for immunity from criminal prosecution for a UK cartel offence, even if their current or former employer-undertaking does not have immunity. The immunity is crystallised by way of a no-action letter issued by the CMA, and is conditional on the individual meeting the same obligations placed on a firm seeking Type A or B immunity, namely that they fully cooperate with the CMA's investigation, have no further involvement in the cartel, and must not at any time have attempted to coerce another undertaking to join the cartel.

No-action letters may be revoked if an individual is found not to have complied with these obligations, or to have knowingly or recklessly provided information that is false or misleading. No-action letters are issued to individuals who, but for having been granted immunity, would ordinarily have been a suspect in the CMA's investigation and subsequently a defendant in any trial. Therefore, not all former and current employees who are covered by conditional immunity will necessarily need a no-action letter.

5.2 Criminal investigations

In the UK, individuals may also be subject to criminal prosecution for certain anti-competitive conduct, by virtue of section 188 of the Enterprise Act 2002, which concerns the criminal cartel offence. Corporate entities cannot be criminally liable for criminal cartel conduct under the 2002 Act. Responsibility for prosecution lies with the CMA and the Serious Fraud Office (SFO), or with a third party subject to the CMA's consent. Investigations are likely to be led by the CMA, in consultation with the SFO if necessary. In Scotland, prosecutions may only be brought by the Crown Office and Procurator Fiscal Service, which is headed by the Lord Advocate.

(a) Offences

An individual is guilty of the cartel offence if he/she agrees with one or more other persons to make or implement, or to cause to be made or implemented, specific kinds of arrangement relating to at least two undertakings. The kinds of arrangement covered are:

- directly fixing a price for the supply of a product or service;
- limiting or preventing the supply of a product or service;
- dividing supply of a product or service with another undertaking;
- dividing customers of a product or service with another undertaking; and
- rigging a competitive bidding process.

Essentially, these arrangements are of either a price-fixing, market-sharing or bid-rigging nature. The cartel offence is not committed where arrangements contain restrictions on only one party, restrictions on more than one party that relate to different parts of the production or supply chain, or do not relate to one of the five kinds of arrangement listed above.[110] At the time that the criminal cartel offence came into being in June 2003, and in order for an individual to be found guilty, it was necessary for the prosecution to prove that the agreement in question was dishonest. Following detailed reconsideration of this element of the offence, the requirement to prove dishonesty was removed from section 188 of the Enterprise Act 2002 on April 1 2014.[111]

There are three exclusions to the cartel offence.

- The 'customer notification' exclusion applies when specified 'relevant information' has been given to customers before entering into a supply arrangement with another an undertaking.
- The 'publication exclusion' applies where relevant information has been published in the London, Edinburgh or Belfast *Gazette* before any arrangements are implemented.

---

110 Competition and Markets Authority, *Cartel Offence Prosecution Guidance*, March 2014.
111 Section 47 of the Enterprise and Regulatory Reform Act 2013.

- The 'bid-rigging notification exclusion' applies where the party requesting bids is given relevant information about the arrangements at or before the time a bid is made.

The 'relevant information' which must be disclosed in all three cases comprises the names of the undertakings to which the arrangements in question relate, a description of the arrangements that demonstrates why they fall, or might fall, within the scope of the offence, and details of the products or services to which the arrangements relate.

(b) ***Defences***

There are three specific defences to the criminal cartel offence.[112]

- The first defence is that there was no intention to conceal the arrangements from customers. This does not specifically require any express disclosure to customers, though this may helpfully demonstrate the point.
- The second defence is that there was no intention to conceal the arrangements from the CMA. In practice, this may be difficult to prove without some form of disclosure.[113]
- The third defence is the legal advice defence. This is that an individual, before entering into or implementing an arrangement that falls within the cartel offence, took reasonable steps to ensure that the nature of the arrangement be disclosed to professional legal advisers for the purpose of obtaining advice about it. In order for the defence to be relied upon, it is not necessary for the individual to prove that he/she followed the advice, and he/she is not required to disclose any legal advice provided. The defence applies even if the legal advice stated that the arrangements were likely to be unlawful and that advice was not subsequently followed.

In view of these defences, an individual seeking to minimise his/her risk of prosecution for the cartel offence should seek legal advice at an early stage and consider disclosing details of any arrangement to customers where it is commercially viable to do so, especially where any form of joint selling or bidding is involved.

(c) ***Criminal penalties***

Under the Enterprise Act 2002,[114] individuals found guilty of the cartel offence are liable on conviction in the Crown Court to imprisonment of up to five years, a fine, or both, and on conviction in the magistrates' court to imprisonment of up to six months, a fine (typically not exceeding £5,000), or both. The magistrates' court can refer a conviction to the Crown Court for sentencing if it considers its sentencing powers to be insufficient.

To date, there have been very few prosecutions for the criminal cartel offence. In

---

112 Section 188B of the Enterprise Act 2002.

113 The CMA is under no duty to respond if it is notified of an arrangement.

114 Section 190 of the Enterprise Act 2002.

the period between the criminal cartel offence coming into existence in 2003 and 2014, when Parliament's removal of the requirement to prove dishonesty in the commission of the offence came into effect, there was only one successful prosecution, against the Marine Hose Cartel,[115] brought by the then Office of Fair Trading. This case involved three British nationals (two of whom pleaded guilty as part of a plea bargain arrangement in the US) who were found guilty of involvement in an international cartel arrangement to fix the global price of marine hose and rig the bids for particular contracts around the world. This was one of only two prosecutions brought during this period. The other was the high-profile prosecution against four British Airways Plc executives (both current and former at the time), which collapsed shortly after coming to trial in 2010, when all four defendants were acquitted.[116] In 2014 and 2015, the CMA prosecuted its first case since the British Airways acquittals, this time in the cylindrical galvanised steel tank industry, in which three executives were charged with market sharing and fixing the price of liquid storage tanks over a seven-year period.[117] Two of the defendants were acquitted following trial and the other pleaded guilty before trial. The charges in this case were brought under the old cartel offence requiring proof of dishonesty, since the alleged conduct predated the change in the law in 2014, and the two defendants were acquitted because the CMA was unable to meet that requirement.

5.3 Private enforcement

Individuals and entities may also use competition law in private litigation, in particular:

- as a defence to an action based on a contract which contains putatively anti-competitive provisions; or
- in an action for damages against a party to anti-competitive conduct.

A term of an agreement that breaches article 101 of the TFEU or the prohibition under Chapter I of the Competition Act 1998 is void and unenforceable.[118] Where an agreement contains such a term, it is for national law to determine whether an agreement containing prohibited terms is invalid altogether, or if the terms in question can be 'severed' from the agreement, so as to leave the rest of it valid.[119] In English law, terms in an agreement may be severed unless that severance would cause the agreement as a whole to fail (eg, for lack of consideration), or to be so changed that it is distinct in character from the agreement the parties originally entered into.[120]

(a) Damages

Individuals and entities may claim for losses suffered as a result of breaches of competition law by virtue of articles 101 and 102 of the TFEU or chapters I and II of

115 *R v Whittle, Allison, & Brammar* [2008] EWCA Crim 2560.
116 *R v George, Crawley, Burns and Burnett* [2010] EWCA Crim 1148.
117 CE/9623/12.
118 See article 101(2) of the TFEU; section 2(4) of the Competition Act 1998.
119 *Société Technique Minière v Maschinenbau Ulm GmbH* [1966] ECR 235.
120 *Chemidus Wavin Ltd v Société pour la Transformation et l'Exploitation des Résines Industrielles SA* [1978] 3 CMLR 514

the Competition Act 1998. As a matter of EU law, a party to an agreement that infringes article 101 may even sue the other contracting party for that infringement unless it bears significant responsibility for the distortion of competition itself.[121]

In order to prove breach, a claimant may rely on any infringement decision against the defendant by the CMA or the Commission, but otherwise will itself have to adduce evidence to prove a violation.[122] The claimant will also have to prove that its loss was caused by the defendant's breach and that the type of loss suffered was reasonably foreseeable. In *Kone AG v ÖBB*,[123] the ECJ held a cartel member liable for an 'umbrella pricing' effect, ie losses suffered by the customer of a competitor which was not part of the cartel but which had raised prices independently as a result of cartel activity.

Several defences are available to claims for losses as a result of breaches of competition law, of which two are key to note here. First, a defendant may argue illegality, on the basis that the claimant bears significant responsibility for the distortion of competition itself and so cannot rely on that illegal action as the basis of a claim. Secondly, under the 'passing-on' defence, a defendant may argue that the damages suffered by the purchaser of its cartelised product are reduced or mitigated if the purchaser has passed on some of the overcharge to its customers.[124]

(b) Jurisdiction

In England & Wales, claimants may bring claims before either the Chancery Division of the High Court or the Competition Appeal Tribunal (CAT). Previously, claims without a prior finding of a violation by the CMA or Commission, known as 'standalone actions', were required to be brought before the High Court. However, after the Consumer Rights Act 2015 (CRA) came into force on October 1 2015, it also became possible to bring standalone actions in the CAT.

'Follow-on' claims, where there has already been a finding of breach by the CMA or Commission, may be brought before either forum.[125] The selection of forum is an important strategic choice to consider in each case, taking account of the different compositions of the High Court and CAT and the different applicable procedural rules.

Claims in respect of breaches of competition law must generally be brought before the High Court within six years of the loss being discovered, although this period is extended if a defendant has deliberately concealed relevant facts.[126] Following the enactment of the CRA, this time limit also applies to claims before the CAT.[127] Previously, follow-on claims before the CAT were subject to a two-year limitation period. Where not all parties to a CMA or Commission decision appeal against it, the time limit for claims against a non-appealing party runs from the date of the original decision. This means that claims against one party to a decision may be barred, although another party's appeal is still ongoing.[128]

---

121 Case C-453/99, *Courage Ltd v Crehan; Crehan v Courage Ltd & Ors* [2001] 3 WLR 1646.
122 See sections 47A(9), 58 and 59 of the Competition Act 1998.
123 Case C-557/12, *Kone AG & Ors v BB-Infrastruktur AG* [2014] 5 CMLR 5
124 See *Fulton Shipping Inc of Panama v Globalia Business Travel SAU* [2014] EWHC 1547 (Comm) for an exploration of the principles underlying the passing-on defence.
125 Schedule 8, Part 1 of the CRA.
126 Section 32(1)(b) of the Limitation Act 1980.
127 Schedule 8, Part 1 of the CRA 2015.

Claims may also be brought by a person representing a number of consumers. Such proceedings may be brought either on behalf of all those who expressly notify the representative that they wish to opt-in or, under reforms introduced by the CRA, on behalf of a defined class with claimants automatically included unless they specifically opt out. These actions may be brought in respect of infringements that have already been discovered and sanctioned by the UK and EU authorities.[129] Where claims for damages brought by an individual are to be continued as a collective action, the individual must expressly consent to this.

The CRA also introduced a fast-track procedure designed for claims by small and medium-sized enterprises in the CAT. Under this procedure, there are restrictions on costs and evidence, at the CAT's discretion, and a CAT chairperson ultimately decides on allocation of the claim. Fast-track cases must in general take no longer than three days and their use is likely to be limited to simple matters.

Under the CRA, newly introduced voluntary redress schemes offer up to a 20% reduction in fines where an undertaking is able to compensate the losses of those affected by its infringements. Approval for a voluntary redress scheme does not protect the undertaking from claims against it, though the CAT will take the scheme into account when deciding whether or not to allow a collective action.[130]

(c) *International choice of forum*

Private enforcement of EU competition law is regulated by the domestic law of the relevant member state. This can pose issues of certainty concerning standing, disclosure, collective action, the standing of indirect purchasers, the burden of proof, calculation of damages, and the availability of the passing-on defence.

Council Regulation (EC) 44/2001 generally requires that persons domiciled or established in a member state be sued in the courts of that state. This is, however, subject to notable exceptions. First, where several defendants are domiciled in different states, they may be sued in the courts of any member state in which one of them is domiciled.[131] Secondly, a person affected by unlawful cartel activity has the option of bringing action against the cartel not only in the state in which the activity occurred but also in the state in which the loss arose,[132] although in the latter case the court's jurisdiction is limited to the loss suffered by the person based in that state.

Jurisdiction clauses will only have effect in relation to litigation for damages arising from an unlawful cartel if the victim in question has consented to the clause. For example, in *Ryanair v Esso*,[133] the Court of Appeal held that a term of the agreement in question which stated that it was to be governed by the laws of

---

128 *Deutsche Bahn AG v Morgan* [2014] UKSC 24.

129 Schedule 8, Part 1(5) of the CRA.

130 Schedule 8, Part 1(12) of the CRA 2015.

131 See Court of Justice of the European Union Press Release 58/15. Further, once that court takes jurisdiction, the withdrawal of the action against the only domiciled defendant does not in principle affect the jurisdiction of the courts with regard to the remaining defendants.

132 Case C-352/13, *Cartel Damage Claims (CDC), Hydrogen Peroxide SA v Akzo Nobel NV & Ors* [2015] All ER (D) 50 (Jun).

133 *Ryanair Ltd v Esso Italiana Srl* [2013] EWCA Civ 1450.

England could not be relied on, as the clause did not envisage competition claims being brought pursuant to a breach of article 101.

Courts in England & Wales have generally been willing to interpret the scope of their jurisdiction generously. For example, in *Toshiba v KME Yorkshire,*[134] they confirmed their jurisdiction over a matter in which none of the addressees of the original Commission decision were domiciled in the UK, given that a UK-based subsidiary was added as a defendant in the private action.[135]

(d) The Damages Directive

Directive 2014/104/EU (the 'Damages Directive') has entered into force and member states must implement it by December 27 2016. The Damages Directive will make it significantly easier to claim compensation for violations of EU competition law and it fine-tunes the interplay between public and private enforcement. In particular, the directive:[136]

- facilitates better access to evidence;
- states that a final infringement decision of a national competition authority will constitute full proof before civil courts in the same member state in which the infringement occurred, and will be *prima facie* evidence that the infringement occurred in other member states;
- establishes clear limitation periods of at least five years;
- codifies the 'passing on' defence;
- clarifies that victims are entitled to full compensation, including actual loss, loss of profit, and interest;
- establishes a rebuttable presumption that cartels cause harm, which also assists in the quantification of loss; and
- establishes the joint and several liability of every participant in an infringement.

(e) Remedies

The measure of damages to which both direct and indirect purchasers of a cartelised product are entitled is full compensation of their losses. This encompasses actual losses, loss of profit, and interest. In the case of an indirect purchaser, he/she must provide evidence of an infringement of competition law, resultant overcharge by the direct purchaser, and purchase of goods from the direct purchaser.

EU courts make use of expert evidence in competition law cases, and economists can play a particularly important role. Expert evidence may be used to calculate the amount a claim is worth, whereby the actual price of the product will be compared with a calculation of what the price would have been but for the product being cartelised. Much of the challenge is in calculating the latter value. In the US, courts

134 *Toshiba Carrier UK Ltd v KME Yorkshire Ltd* [2012] EWCA Civ 169.

135 As a result, the claim became mixed as opposed to purely follow-on.

136 European Commission, *Competition Policy Brief – The Damages Directive*, January 2015.

137 Named after the first in a line of three cases that articulate the *Daubert* test – *Daubert v Merrell Dow Pharmaceuticals* [1993] 509 US 579, *General Electric Co v Joiner* [1997] 522 US 136, and *Kumho Tire Co v Carmichael* [1999] 526 US 137.

apply the *Daubert* test to test economic evidence This test requires that the evidence be scientifically valid and focus solely on principles and methodology.[137]

Exemplary damages are in principle available in English law, at the court's discretion, for infringements of competition law in cases where it is deemed necessary to punish and deter the defendant.[138] Where, however, fines have already been imposed on a defendant by an EU or UK competition authority, or would have been imposed but for a successful leniency application, exemplary damages will not be awarded.[139] In *Albion Water v Dŵr Cymru,*[140] however, the CAT held that the claimant's claim for exemplary damages should not be struck out, in spite of there being an infringement decision against the defendant, because no fine had been imposed pursuant to the decision and there was no practical possibility of one being imposed by an enforcement body in the future, so that there was no risk of duplication of penalty or double-counting. Courts are less likely to award exemplary damages where there are multiple claimants or potential claimants, given the difficulty of apportioning such damages.

(f) ***Injunctions and interim relief***

The High Court may grant an injunction to restrain a breach of competition law pending trial, applying a flexible modified *American Cyanamid* approach.[141] The CAT previously followed the test used by the General Court of the EU, which requires the party seeking relief to demonstrate that it will otherwise suffer serious and irreparable harm. However, in *Sky v Ofcom,*[142] which was eventually resolved by consent so no judgment was given, CAT President Barling J indicated that he was minded to adopt an approach analogous to that used by the High Court, rather than the CAT's previous approach.[143]

(g) ***Arbitration***

Competition law issues may be referred to arbitration, in particular disputes as to the enforceability of terms. The Commission has, however, been known to investigate cases after parties have settled their dispute in arbitration proceedings. Arbitrators cannot refer issues of competition law to the ECJ under article 234 of the TFEU, given that the establishment of an arbitral tribunal is a consensual process resulting in something that is not a court or tribunal of a member state.[144] Further, there is no obligation on an arbitral tribunal root out infringements of competition rules actively.[145]

---

138 *Kuddus v Chief Constable of Leicestershire* [2002] 2 AC 122, per Lord Nicholls.

139 *Devenish Nutrition Ltd v Sanofi-Aventis SA and Ors* [2007] EWHC 2394 (Ch).

140 *Albion Water Ltd v D r Cymru Cyfyngedig* [2010] CAT 30.

141 *American Cyanamid Co v Ethicon Ltd* [1975] AC 396. See CPR 15-7 & 15-8, *Civil Procedure Rules 2015* Volume 2 (2015],(London: Sweet & Maxwell) for further detail.

142 *British Sky Broadcasting Ltd v Office of Communications,* CAT, 29 April 2010.

143 Dinah Rose QC & Tom Richards, *Appeal and Review in the Competition Appeal Tribunal and High Court* (Blackstone Chambers, 2010).

144 Case C-102/81, *Nordsee Deutsche Hochseefischerei GmbH v Reederei Mond Hochseefischerei Nordstern AG & Co KG* [1982] ECR 1095.

145 Joined Cases C-430/93 and C-431/93, *Van Schijndel and van Veen v Stichting Pensioenfonds voor Fysiotherapeuten* [1995] ECR I-4705.

In view of the time pressures often inherent in agreements for the supply of energy, there is occasionally a need for urgent interim relief before substantive arbitration proceedings can take place. Traditionally, parties in these circumstances would have to look to the courts at the seat of the arbitration for relief. This may not always be desirable, potentially for the same reasons that led arbitration to be chosen as an alternative to begin with. Increasingly, the procedural rules of arbitral institutions have made provision for parties to seek urgent interim relief through emergency arbitration. Questions remain, however, as to the standing of so-called 'emergency arbitrators' and the enforceability of their decisions.

Sanctions

Hannah Laming
Peters & Peters Solicitors LLP
Rebecca Wright
23 Essex Street Chambers

1. Introduction

Sanctions are imposed by international organisations and individual countries and are frequently used as a political tool, to encourage change in a particular country or regime where diplomatic efforts have failed. There are various rationales for imposing sanctions – for example, to deal with contraventions of international law, the suppression of terrorism and terrorist financing, other threats to national or international security or as a reaction to human rights abuses perpetrated by the state. They can also be implemented for domestic reasons, for example, to protect a nation's commercial interests.

Economic sanctions prohibit or restrict trade and financial transactions with individuals or entities.

The oil and gas industry is regularly the target of sanctions due to the powerful impact this sector has on the economy and on a country's ability to develop nuclear programmes. In Iran, sanctions on oil exports resulted in Iran's crude oil sales revenue falling from $100 billion in 2011 to about $35 billion in 2013.[1]

Complying with international sanctions is challenging, particularly for multinational corporations that may be subject to numerous different and sometimes incongruent sanctions regimes. The complex nature of sanctions is exacerbated by the extra-territorial effect of some sanctions programmes, the international nature of trade and finance and the fact that, as sanctions respond to political developments across the globe, they are in a state of constant flux.

Sanctions compliance is an area of growing concern for companies operating in the oil and gas sector, as well as those providing goods or services to such companies, particularly given the serious impact a breach of sanctions can have on a company, both financially and reputationally.

This chapter gives an overview of how sanctions work, focusing on the European Union (EU) and United States (US) sanctions regimes. The authors explore different types of sanctions and their impact on the oil and gas industry, address issues of interest to multinational corporations and identify the critical components of an effective compliance programme.

As sanctions react to geopolitical concerns, they are continually developing and it is beyond the scope of this chapter to identify all sanctions worldwide relating to the oil and gas sector. While this chapter provides a good starting point for

1 Kenneth Katzman, *Iran Sanctions*, Congressional Research Service (April 21 2015).

understanding sanctions and the legal position is correct at the time of writing, anyone engaging in business activity in that field should ensure that they familiarise themselves with all relevant sanctions that might impact upon their business.

2. Overview of international sanctions programmes

Many international organisations and countries implement sanctions regimes. This section gives an overview of some of the principal sanctions regimes and how they are implemented.

The United Nations (UN) Security Council imposes sanctions pursuant to article 41 of the United Nations Charter ('the Charter'), which enables the Security Council to enforce its decisions using methods not involving armed forces – for example, the interruption of economic relations. Members of the UN are obliged to implement UN sanctions under article 25 of the Charter, but the UN does not prescribe methods of implementation. It is common for member states to adopt a single piece of enabling legislation to facilitate the transposition of UN sanctions into domestic law (for example, through regulations). Some members, however, implement UN resolutions on a case-by-case basis. The various methods of transposing UN resolutions into domestic law can also give rise to variations in the nature and scope of the different domestic sanctions regimes.

The EU adopts UN sanctions by adopting a Common Position and then, generally, by implementing an EU Directive. EU Directives have direct effect in member states but will usually allow each member state some discretion in their detailed implementation in national law.

The EU also imposes its own sanctions independently of the UN.

Various other bodies and organisations around the world also have sanctions programmes, for example, the Arab League. In addition, individual countries may operate their own sanctions regimes on the basis of their national interests (most notably, the US).

3. Scope of application

When considering to whom a sanctions regime applies, it is necessary to consider the specific provisions of that regime. The scope of application varies widely between different sanctions programmes and different regimes within the same jurisdiction may have different application. The following is an outline of who is affected by EU and US sanctions, as well as liability for the action of a subsidiary under these regimes.

3.1 EU sanctions

EU sanctions apply to:

- any EU national, wherever located. EU nationals living or working overseas must not engage in any activities or transactions on their own behalf that would breach EU sanctions, but are permitted to engage in such transactions if they are conducted in the course of business for a company outside the EU that is not subject to EU sanctions;
- any entity incorporated or constituted under the law of an EU member state, for example, the branches or wider representative offices of EU incorporated entities;

- any individual or entity doing business within the EU. This includes non-EU nationals working for EU companies and the activities of non-EU companies that take place wholly or partly in the EU.

3.2 US sanctions

US sanctions apply to all 'US persons', which include:

- US citizens and permanent resident aliens,[2] wherever they are located. This includes any US citizen or permanent resident working outside the US for a non-US entity. US citizens and US permanent resident aliens are not permitted to engage in any activity or transaction that would breach US sanctions, even if it is performed in the course of business for an entity not subject to US sanctions;
- persons within the US regardless of their nationality. This includes non-US citizens visiting the US or working for US-based entities; and
- companies incorporated under the laws of the US or of any state, territory or possession of the US. This includes foreign entities.

In addition, the US often has jurisdiction, particularly in relation to trade embargoes, where there is any US nexus to a transaction (ie, it can take enforcement action against non-US institutions, provided that it can prove a US nexus). The US will deem there to be a nexus if a US individual or entity is involved at any stage of the transaction. This includes the use of US dollars (because they are cleared through financial institutions in the US), the use of US financial institutions or insurance companies, or knowingly relying on services provided by US persons anywhere in the world to facilitate, participate in, approve or support restricted transactions.

Finally, some US sanctions legislation makes it an offence for anyone, wherever they are located, to breach certain sanctions. Sanctions on non-US persons operating entirely outside of US jurisdiction are called 'secondary sanctions'. For example, prior to sanctions relief granted to Iran under the Joint Comprehensive Plan of Action ('JCPOA') when it was implemented in January 2016,[3] the US Comprehensive Iran Sanctions, Accountability and Divestment Act ('CISADA') made it an offence for any person (individual or entity) to knowingly invest $20 million or more in Iran's energy sector.[4] In 2011, the US government found that Belarusneft, a state-owned energy company in Belarus, had breached this provision when it invested $500 million in a contract with the Iranian energy company Naftiran Intertrade.[5] As a consequence, the US government prohibited Belarusneft from receiving export assistance from the US Export-Import Bank, obtaining export licences or private US bank loans exceeding $10 million per year, or securing any procurement contracts

---

2 This includes individuals who are not US citizens but have a permanent right to live in the US (eg, green card holders).

3 US Department of State, *Joint Comprehensive Plan of Action* (February 1 2016), available at www.state.gov/e/eb/tfs/spi/iran/jcpoa.

4 Comprehensive Iran Sanctions, Accountability and Divestment Act 2010, 22 USC 8501 (July 1 2010), section 102(a)(1)(a)(1).

5 See US Department of State, *Iran Sanctions Act Announcement* (March 29 2011), available at www.state.gov/r/pa/prs/ps/2011/03/159309.htm.

from the US government.[6] These sanctions were imposed despite the fact that Belarusneft did not operate in, or have any links to, the US.

3.3 EU attempts to limit the extra-territorial effect of US sanctions

The EU has responded to the US adoption of sanctions with extra-territorial scope by enacting legislation prohibiting compliance with them. Council Regulation (EC) 2271/96 prohibits EU companies or individuals (or those in the EU) from complying with specific US extra-territorial legislation, including some of the US sanctions regulations.[7]

3.4 Liability for actions of subsidiaries

In the EU, the question of whether a parent company can be held liable for sanctions breaches committed by one of its subsidiaries or a related company that is not itself subject to EU sanctions regulations is assessed on a case-by-case basis. There is no principle of 'vicarious liability' that would render an EU company liable solely by virtue of a breach committed by its subsidiary.

Whether the parent company is held liable would depend, for example, on the extent to which the subsidiary is a separate entity or acting as an agent of the parent company. It would need to be shown that the EU parent company had itself been involved in any breach.

US sanctions are generally more precise on the question of liability for breaches by a subsidiary. US sanctions often extend to foreign entities "owned and controlled" by a US parent. According to US Iranian sanctions, a foreign entity is "owned and controlled" by a US person if the US person:

> *a) holds a 50% or greater equity interest by vote or value in the entity; b) holds a majority of seats on the board of directors of the entity; or c) otherwise controls the actions, policies or personnel decisions of the entity.*[8]

In November 2013, Weatherford International Ltd (Weatherford), a Swiss oil services company listed on the New York Stock Exchange, and four of its subsidiaries located in Canada, the United Arab Emirates and the United Kingdom, admitted exporting oil and gas equipment to Iran, Syria and Cuba between 1998 and 2007 in breach of US sanctions. Weatherford entered into a deferred prosecution agreement and the US government imposed a combined total penalty of $100 million, which included civil and criminal penalties.[9]

4. Designated persons

Many sanctions regimes prohibit transactions with specified individuals and entities ('Designated Person(s)'). The EU publishes a *Consolidated List* of persons, groups and entities subject to EU financial sanctions.[10] In the UK, the list of individuals subject

---

6 Daniel Seleanu, US sanctions Belarus state-owned oil company for $500m Iran investment (March 31 2011), available at www.complinet.com/global/news/news/article.html?ref=142633.

7 *Ibid*, article 5.

8 Iranian Transactions and Sanctions Regulations, CFR 560.215(b)(1).

9 Department of Justice, *Three Subsidiaries of Weatherford International Limited Agree to Plead Guilty* (November 26 2013), available at www.justice.gov/opa/pr/three-subsidiaries-weatherford-international-limited-agree-plead-guilty-fcpa-and-export.

to sanctions ('Designated Persons') is contained in Her Majesty's Treasury (HMT) *Consolidated List of Targets*.[11] In the US, the equivalent list is called the *Specially Designated Nationals List* ('SDN').[12]

4.1 European Union

Under EU law, a person or entity commits an offence if, without a licence, they provide funds or economic resources where they know or have reasonable cause to suspect that they are:

- making funds or economic resources available directly or indirectly to a Designated Person;
- making funds and economic resources available to a third party for the benefit of a Designated Person.[13]

The indirect provision of funds relates to a situation whereby funds or economic resources are provided to a Designated Person via a non-designated third party.

There is a presumption that making funds or economic resources available to non-listed persons owned or controlled by a Designated Person or entity will constitute an indirect provision to a Designated Person unless it can be shown that the funds or economic resources concerned will not be used by or be for the benefit of the Designated Person.[14] The relevant criteria to be taken into account include:

- the date and nature of the contractual links between the entities concerned (for example, sales, purchase or distribution contracts);
- the relevance of the sector of activity of the non-listed entity for the listed entity; and
- the characteristics of the funds or economic resources made available, including their potential practical use by, and ease of transfer to, the listed entity.

An entity is owned by another person or entity where they possess more than 50% of the proprietary rights or have a majority interest.[15] Determining whether an entity is controlled by another person or entity is more complex. While there is no formal guidance on this point, proposed European Council Guidance issued in April 2013[16] identifies a number of possible criteria, which include:

- the right to appoint or remove a majority of the members of the administrative, management or supervisory body of the entity;
- controlling, alone or with others, a majority of shareholders' or members' voting rights;

10 See http://eeas.europa.eu/cfsp/sanctions/consol-list/index_en.htm.

11 HMT, *Consolidated List of Financial Sanctions Targets in the UK*, available at http://hmt-sanctions.s3.amazonaws.com/sanctionsconlist.htm.

12 Office of Foreign Assets Control (OFAC), *Specially Designated Nationals List*, available at www.treasury.gov/resource-center/sanctions/SDN-List/Pages/default.aspx.

13 See, for example, article 23(3) of EU Regulation 267/2012 of 26 March 2012 concerning restrictive measures against Iran).

14 Council of the EU, Guidelines on implementation and evaluation of restrictive measures (sanctions) in the framework of the EU Common Foreign and Security Policy (April 30 2013), paragraph 4.

15 *Ibid*, paragraph 1.

16 *Ibid*, paragraph 2.

- the right to exercise a dominant influence over a legal person or entity;
- the right to use all or part of the assets of the entity;
- managing the business of a legal entity on a unified basis, while publishing consolidated accounts;
- sharing jointly and severally the financial liabilities of the entity or guaranteeing them.

Providing an economic resource to a non-listed person to generate profits which might, in part, be distributed to a listed shareholder will not, however, constitute a breach of the EU provisions.[17]

The EU has listed a number of individuals and entities as Designated Persons because of their involvement in the oil and gas sector. These designations principally relate to entities in Syria and, prior to implementation of the JCPOA, also related to entities in Iran. There are some EU-listed entities in Iraq, but the restrictions only apply to economic resources received before May 23 2003. There are a few listed entities in the Ukraine.

Entities designated by the EU include:

- many oil, gas and petroleum companies and their subsidiaries, including the Syrian Petroleum Company;
- banks and financial centres connected with payments relating to oil and gas industries;
- organisations trading equipment for the oil and gas sector;
- companies involved in shipping oil or as intermediaries supplying oil;
- organisations providing financing and expertise to Iran's petroleum sector; and
- individuals linked to oil, gas and petroleum companies or activities, for financing and facilitating transactions in Iranian crude oil.

4.2 United States

Similar restrictions apply to the provision of economic resources to Designated Persons in the US.

Under US sanctions, there are entities which are deemed to be blocked even though they do not appear on the SDN List because they are owned or controlled by a Designated Person. In a similar manner to the EU, an entity is deemed to be blocked where a Designated Person owns (directly or indirectly) a 50% or greater interest. For that reason, some OFAC sanctions programmes (for example, those targeting Cuba and Sudan), contain a broader category of entities who are blocked because of issues around control and ownership.[18]

Frequently, individuals and entities designated by the EU will also be designated by the US. The US SDN List is, however, generally more comprehensive and this is true in relation to the oil and gas industry. A good example of this is the more extensive list of designated individuals and entities linked to the Ukrainian oil and gas industry.

17 *Ibid*, paragraph 4.
18 www.treasury.gov/resource-center/sanctions/Documents/licensing_guidance.pdf.

The US SDN List currently includes:

- entities involved in the exploration, production, transportation, storage and processing of gas and oil, eg, Rosneft (the world's largest traded oil company), Gazprom and Transoil;
- construction companies undertaking projects in the energy sector, eg, Stroytransgaz Group;
- private entities involved in the transport of oil and petroleum products, eg, Transoil;[19] and
- individuals linked to the oil and gas industry, eg, Igor Sechin, the chairman of Rosneft.

4.3 Screening programme

A standard control implemented by organisations to mitigate the risk that they will provide funds or economic resources to a Designated Person is a screening programme which screens the names of those individuals and entities with which it transacts against relevant Sanctioned Party Lists.

To ensure a screening programme is effective, it is important to ensure:

- accurate inputting of data;
- sufficient due diligence is carried out to identify situations where entities are not themselves designated but are owned or controlled by a Designated Person; and
- that the screening process is appropriate in terms of scope, timing and frequency.

4.4 Applying for a licence

In limited circumstances, sanctions programmes will permit economic resources to be made available to a Designated Person provided that a licence is obtained in advance from the relevant authority (eg, OFAC in the US or HMT in the UK). There is no centralised body that can issue licences across the EU, so that applications must be made to the relevant member state(s).

4.5 Legal challenges to designation under sanctions

The decision to list a particular entity or individual as a Designated Person can be legally challenged. For example, in July 2014 the National Iranian Tanker Company (NITC), an Iranian company that operates one of the world's largest fleets of double-hulled tankers for the transport of oil and gas, had its designation on European sanctions measures annulled by the European Court of Justice.[20] NITC claimed that the reasons stated in the contested decision and regulation were vague and did not satisfy the requirement that the reasons given must be specific and concrete.[21] The

---

19 OFAC, *Ukraine-related Designations* (April 28 2014), available at www.treasury.gov/resource-center/sanctions/OFAC-Enforcement/Pages/20140428.aspx.

20 *National Iranian Tanker Company v Council of the European Union*, European Court of Justice (Seventh Chamber), Case T-565/12 (July 3 2014), available at http://curia.europa.eu/juris/document/document.jsf?docid=154524&doclang=EN.

21 *Ibid*, paragraph 32.

court agreed that that there were insufficient reasons for the listing and held that the material provided for consideration "contains no evidence capable of supporting the Council's claims that the applicant was controlled by the Government of Iran and provided financial support to the Government of Iran."[22]

5. Country- and sector-specific sanctions

5.1 Trade embargoes and targeted sanctions

In addition to sanctions put in place against designated individuals and entities, sanctions are often imposed to prohibit or restrict trade with a specific country or countries (the 'Sanctioned Country' or 'Sanctioned Countries').

The most restrictive sanctions are trade embargoes, which effectively prohibit any trade with a country, subject to very limited exceptions (eg, the provision of humanitarian aid). Trade embargoes are typically imposed by the US and Canada. The US currently has trade embargoes in relation to Iran, North Korea, Sudan, Syria and Cuba. Recently, however, the US has relaxed the Cuban trade embargo slightly and, subject to conditions, has carved out certain exceptions to its Iran embargo, particularly for non-US persons, including relief for the oil and gas sector (see below). Examples of activities typically prohibited by US trade embargoes include:

- the importation of goods or services from the Sanctioned Country into the US (or a US territory) or by a US person (including all US citizens and US-registered firms abroad);
- the export, transhipment or re-export of goods or services from the Sanctioned Country or originating from the Sanctioned Country from the US or by a US person;
- the performance of any contract which relates to the Sanctioned Country by any US person;
- facilitating transactions connected to the Sanctioned Country (eg, the provision of financial services); and
- dealing in property in which the Sanctioned Country or a national of the Sanctioned Country has an interest.

Targeted sanctions can relate to specific activities or sectors and frequently encompass the oil and gas sectors. Some countries and international organisations (including the US and the EU) also impose import or, more commonly, export controls in relation to specified products or services originating from Sanctioned Countries. These sanctions generally require individuals or entities to obtain a licence from a central body (for example, OFAC in the US or HMT in the UK) before they can import or export the relevant goods or services.

The EU's export controls include a prohibition on the export of 'Dual-Use Items', which are defined by the UK government as "those that can be used for civil or

22 *Ibid*, paragraph 64.
23 Gov.UK, Do I need an Export Licence? (September19 2013), available at https://www.gov.uk/beginners-guide-to-export-controls.

military purposes – which meet certain specified technical standards and some of their components",[23] but its sanctions regimes have also imposed, for example, prohibitions on trade relating to luxury goods,[24] precious metals[25] and, frequently, the oil and gas sector (see below).

5.2 Examples of sanctions regimes by country

(a) *Iran*

Iran is currently subject to sanctions regimes imposed by both the US and the EU in respect of its nuclear, ballistic missile and other WMD programmes, alleged support for terrorism, and human rights record. US and EU nuclear-related sanctions on Iran have changed substantially following the implementation of the JCPOA in January 2016, including by easing restrictions applicable to the country's oil and gas sectors.

Prior to these changes, both the US and EU regimes imposed prohibitions on importing Iranian oil or gas and related products,[26] financing,[27] investing in[28] or providing key equipment[29] and technical assistance[30] to the oil and gas sector in Iran, and providing transport for oil and gas products connected with Iran or an Iranian person.[31] EU sanctions also restricted the making and receiving of payments between financial institutions in member states and those in Iran[32] as well as requiring payments above a certain threshold to and from any "Iranian Person"[33] to be notified to or authorised by the relevant body in the member state.[34] The US prohibitions were part of a regime imposing a near-total ban on US persons transacting with Iran,[35,36] which also imposed restrictions on non-US persons outside of US jurisdiction, as noted above.[37]

Following implementation of the JCPOA, US secondary sanctions no longer apply to (among other things) purchasing or acquiring Iranian petroleum or related products, selling refined petroleum products to Iran, providing goods or services for use in connection with the Iranian energy sector, investing in the Iranian petroleum sector or participating in joint ventures concerning petroleum with Iran outside of the country, and assisting the National Iranian Oil Company and Naftiran Intertrade Company. Likewise, the EU has lifted prohibitions on dealing in Iranian gas and petrochemical products, providing goods and services for use in the oil and gas sector in Iran, and investing in or entering into joint ventures relating to that sector. In

---

24 Council Regulation (EU) 36/2012, article 11b.
25 *Ibid*, article 11a.
26 Council Regulation (EU) 267/2012, article 11.
27 *Ibid*, article 9.
28 *Ibid*, article 17.
29 *Ibid*, article 8.
30 *Ibid*, article 9.
31 *Ibid*, article 37b.
32 *Ibid*, article 30.
33 An individual or entity in or resident in Iran.
34 *Ibid*, article 30a.
35 Executive Order 12613 (Imports) & Executive Order 13059 (Exports).
36 Iranian Transactions and Sanctions Regulations, 41 USC § 560 (January 28 2016).
37 Comprehensive Iran Sanctions, Accountability, and Divestment Act 2010, 22 USC § 8501 (July 1 2010), section 102(a)(1)(a)(1).

addition, the EU lifted asset freezes against and the US removed from its SDN list several companies involved in the Iranian oil and gas sector.

The US continues to enforce its general prohibition on US persons doing business with Iran, restrictions on use of the US financial sector in connection with Iran, and several secondary sections, and all US and EU sanctions that have been lifted are subject to being 'snapped back' into force should Iran fail to comply with its obligations under the JCPOA.

The sanctions on Iran have had a significant impact on its oil and gas sector. For example, following the EU's ban on importing Iranian crude oil in 2012,[38] the country's crude oil exports fell to approximately 1.2m barrels per day at the end of 2013, from approximately 2.5m barrels per day in 2011.[39]

(b) Russia

Following Russia's annexation of Crimea and Sevastopol from Ukraine in March 2014, both the US and the EU imposed sanctions on Russia and the Crimea and Sevastopol regions of Ukraine. The sanctions target Russia's oil and gas industry by prohibiting the provision of goods, services and technology for use in important areas of new development, such as exploration or production in connection with deep water, Arctic offshore or shale projects.[40,41] The sanctions also significantly curtail the access of Russian oil companies Rosneft, Transneft and Gazprom Neft, as well as gas companies Novatek and Gazprom under the US regime only, to financial markets and liquidity.[42,43] Combined, these sanctions make it more challenging for Russia's oil sector, and to a lesser extent its gas sector, to acquire the necessary equipment and raise sufficient capital to undertake new projects.

Although the sanctions have impacted upon new developments in Russia's oil and gas industry, major European oil companies have continued to sign deals with Russian counterparts in anticipation of the sanctions eventually being lifted.[44] In addition, the fact that the EU allows contracts agreed before sanctions were introduced to continue, irrespective of the prohibitions, lessens their force.[45] Furthermore, Europe's reliance on Russian gas, the latter accounting for around 30-40% of the region's total supply,[46] has so far discouraged the EU from following in the footsteps of the US by targeting Russian gas companies alongside oil companies.

(c) Syria

Syria is currently subject to wide-ranging US and EU sanctions in response to the

---

38 Council Regulation 267/2012, article 11.

39 Organization of the Petroleum Exporting Countries (www.opec.org/opec_web/static_files_project/media/downloads/publications/ASB2015.pdf).

40 Directive 4 under Executive Order 13662.

41 Council Regulation 833/2014, articles 3 and 3a.

42 Directive 2 as amended under Executive Order 13662.

43 Council Regulation (EU) 833/2014, article 5(2).

44 *EU's Russia sanctions fail to dent oil deals*, Financial Times June 14 2015, available at www.ft.com/cms/s/0/21d66e58-10ef-11e5-8413-00144feabdc0.html#axzz3hSZzaroO.

45 Council Regulation 833/2014, articles 3 and 3a.

46 Eurostat, *Main Origin of Primary Energy Imports, EU-28, 2003–13*, available at http://ec.europa.eu/eurostat/web/energy/data/database.

Syrian government's repression of its civilian population and its violations of human rights. The sanctions target Syria's oil sector by prohibiting virtually all transactions involving Syrian crude oil or petroleum products, or related to the Syrian oil industry generally[47,48]. The sanctions also impose prohibitions in connected areas, such as a general export ban on US goods and services to Syria[49] and a specific EU export ban on particular goods, software and services used in the oil and gas industry and on providing credit to any Syrian individual or entity involved in the oil sector.[50]

Prior to sanctions prohibiting the import of Syrian oil entering force in 2011,[51] the EU was Syria's largest oil export market, purchasing around 90% of all Syrian oil. Although imports of Syrian oil accounted for only 1.35% of Europe's total oil imports in 2010, the exports accounted for 30% ($4.1bn) of Syrian government revenues in that year.[52] The case of Syria illustrates not only the power of oil and gas sanctions to target the economy of states that are reliant on revenue generated by that industry, but also the limitations of such a policy given that, even in view of the importance of the oil and gas sector to Syria, the sanctions have yet to achieve the political goal for which they were designed.

(d) *Libya*

The current US and EU sanctions regimes on Libya were first imposed in accordance with UN Security Council Resolution 1970 of 2011 and initially targeted the now deposed Qadhafi regime. Following a further UN resolution,[53] the focus of those regimes has shifted: instead they target (among other things) illicit exports of oil from the country. In accordance with this resolution, the US and EU authorise the inspection of vessels that the UN has designated and prohibit those vessels from loading, carrying or discharging Libyan crude oil.[54,55]

While US and EU sanctions relating to the oil and gas industry have typically been designed to impact upon the economy of a given state, with the intention of effecting a change in the behaviour of its government, the case of sanctions on Libya demonstrates an alternative way in which they can directly affect the industry. The sanctions are in place to aid Libya's political transition and prevent the embezzlement of its natural resources.

6. Circumvention and facilitation

EU sanctions frequently contain anti-circumvention provisions. These prohibit individuals and entities from engaging in activities which intend to or do circumvent EU sanctions.[56] Provisions of this nature are directed at artificial or sham transactions

47 Executive Order 13582, section 2.
48 Council Regulation (EU) 36/2012, article 6.
49 Executive Order 13338, section 1 and Executive Order 13582, section 2.
50 Council Regulation (EU) 36/2012, article 13.
51 Council Regulation (EU) 878/2011.
52 US Energy Information Administration,. Over 90% of Syrian crude oil exports go to European countries, (September 16 2011), available at www.eia.gov/todayinenergy/detail.cfm?id=3110.
53 UN Security Council Resolution 2146 (2014).
54 *Ibid*, paragraph 10.
55 Council Regulation (EU) 204/2011, article 10b.
56 See, for example, EU Regulation 267/2012, article 41.

designed deliberately to evade the sanctions regime by, for example, using front companies or other concealment methods.

Under US sanctions, it is prohibited for a US Person to approve, finance, facilitate or guarantee any transaction by a foreign person where, if the US Person itself/himself had performed the transaction, that person would have breached sanctions. The expression 'facilitation' could capture, for example, the provision of services by a US insurer or bank in relation to a transaction between third parties. It could also occur where a US citizen on the board of a non-US company is involved in or approved actions that furthered trade or financial transactions with a sanctioned jurisdiction.

An entity that instructs a third party to conduct due diligence on its behalf may remain exposed to liability for any deliberate or negligent act or omission by that third party where the entity breaches a sanction in reliance on that act or omission. Although the vast majority of EU sanctions provide a defence to infringement, it is limited in its scope: an entity can only escape liability where it can prove that it did not know and had no reasonable cause to suspect that the third party's due diligence was deficient.[57]

7. Joint ventures

Joint venture agreements are common in the oil and gas sector. This can be the only way to enter a market where foreign companies are not permitted to establish wholly-owned subsidiaries. The impact of sanctions on joint venture agreements can, however, be sudden and significant. For example, when the EU and the US imposed sanctions against Russia, Western oil companies were forced to discontinue their operations in that country. In September 2014, ExxonMobil stopped offshore Arctic drilling with its Russian partner, Rosneft.[58] Shortly afterwards, Total aborted plans to explore for shale oil in Siberia with its Russian partner, Lukoil.[59] In October 2014, Shell suspended its activities in Siberia with Gazprom Neft.[60] Entities entering into joint ventures should assess the risk of sanctions being imposed and how they would deal with that eventuality from a legal and financial perspective.

8. Insurance and reinsurance

Sanctions can make it difficult for companies operating in the oil and gas industry to obtain insurance or renew an existing policy which relates to a country subject to sanctions. For example, prior to the implementation of the JCPOA, EU Regulation 267/2012 prohibited the provision of "insurance and reinsurance related to the import, purchase or transport of petrochemical products of Iranian origin or that have been imported from Iran".[61]

---

57 See, for example, Council Regulation (EU) 36/2012, article 28.

58 *Exxon, Rosneft said to halt Arctic drilling on sanctions*, WorldOil (September 19 2014), available at www.worldoil.com/news/2014/9/19/exxon-rosneft-said-to-halt-arctic-drilling-on-sanctions.

59 "Sanctions scupper Total/Lukoil venture," Financial Times (September 22 2014), www.ft.com/cms/s/0/49fba004-424d-11e4-a9f4-00144feabdc0.html#axzz3eqAJwBj7.

60 Andy Tully, *Russian Shale Boom Unlikely As Sanctions Force Another Company Out*, Oil Price (October 8 2014), available at http://oilprice.com/Energy/Crude-Oil/Russian-Shale-Boom-Unlikely-As-Sanctions-Force-Another-Company-Out.html.

61 EU Regulation 267/2012, article 13.

9. Contractual clauses

Companies in the oil and gas sector should consider what clauses they can include in contracts with third parties to mitigate sanctions risks. It is increasingly common, for example, to include OFAC compliance clauses in contracts. These clauses can require the other party to the contract to guarantee that they (and all their subsidiaries) are not owned and controlled by a Designated Person and do not undertake transactions in sanctioned jurisdictions. Such clauses do not, however, act as a complete bar to liability. Companies would still be expected to undertake reasonable due diligence on their counterparties and the nature of the transactions involved.

Most contracts contain general *force majeure* provisions that allocate each party's obligations in the circumstances where an event outside a party's control causes contractual performance to be impossible. These can offer protection from liability for breaches arising from the imposition of sanctions.

For example, the English Court of Appeal held, in *Arash Shipping Enterprises Co Ltd v Groupama Transport*,[62] that a broadly-worded *force majeure* clause could entitle a party to cease performing pre-existing contractual obligations due to the risk of breaching sanctions.

Although contractual clauses can serve an important role in allocating risk for sanctions breaches, EU sanctions regulations frequently allow national authorities to issue licences that authorise payments to sanctioned entities or individuals that are due under pre-existing contractual relationships, on the condition that such payments are subsequently frozen. This can resolve issues where companies are otherwise unable to make payments as a result of sanctions.

10. Enforcement and penalties for breaching sanctions

The exact nature and extent of penalties for breaching sanctions depends on the specific provisions in domestic legislation. Individuals and entities can, however, face civil and criminal enforcement action, large fines, reputational damage and, for individuals, the prospect of imprisonment.

11. Corporate compliance programmes

Given the complexity of sanctions generally and the frequency with which they focus on the oil and gas sector, it is imperative for entities operating in markets and/or activities which may engage sanctions to put in place a comprehensive risk-based compliance programme to ensure that they do not breach any sanctions provisions. An effective compliance programme will not only mitigate the risk of breach in the first place, but will also reduce the risk that an entity will be prosecuted for any breach that may in fact occur, as authorities will take into account the adequacy of the proactive compliance steps taken by companies.[63]

A corporate compliance programme should:

---

62 [2011] EWCA Civ 620.

63 See, for example, OFAC, 31 CFR Part 501, *Economic Sanctions Enforcement Guidelines*,, General Factor E, Federal Register (November 9 2009), at 57603, available at www.treasury.gov/resource-center/sanctions/Documents/fr74_57593.pdf.

- be based on a risk assessment for the entity which includes assessing the geographical locations in which it operates, the nature of its activities, any products imported or exported by it, the profile of the entity's customers or suppliers and the nationality of its employees;
- establishing policies, procedures, systems and controls as required to address any sanctions issues identified;
- include appropriate due diligence processes and a screening programme;
- ensure that contractual terms are included as appropriate to avoid or mitigate sanctions risks;
- ensure that employees are trained regularly and appropriately;
- ensure that there is ongoing monitoring and review of the compliance programme in relation to sanctions and that it continues to be effective;
- ensure that there is a culture of compliance within the entity, driven from the top.

Recent proceedings against Schlumberger Oilfield Holdings Ltd (SOHL) demonstrate the importance of multinational companies not only having in place a sanctions compliance programme but also ensuring that this programme operates effectively throughout the business.

On March 25 2015, SOHL admitted that it had wilfully breached US sanctions law after it engaged in illegal transactions and conducted business with Iran and Sudan. The offending actions had been undertaken by Drilling and Measurements (D&M), a Schlumberger business segment in the US. D&M violated sanctions by:

- disguising the company's capital expenditure requests from Iran and Sudan for the manufacture of new oilfield drilling tools and for the spending of money for certain company purchases (eg, subsidiaries outside the US would refer to Iran and Sudan as 'Northern Gulf' and Sudan as 'South Egypt' when seeking approval for transactions);
- making and implementing decisions specifically related to business in Iran and Sudan;
- providing technical services and expertise relating to drilling equipment held in Iran and Sudan.[64]

SOHL did have policies and procedures in place to ensure that D&M did not violate US sanctions, but failed to provide adequate training to its employees to ensure that all US persons, including non-US citizens resident in the US while employed by D&M, complied with Schlumberger's global sanctions policies.

The plea agreement reached with the US government required SOHL to pay $232,708,356 and enter into a three-year corporate probation period, during which time it had to maintain its cessation of all operations in Iran and Sudan.

64 Department of Justice, *Schlumberger Oilfield Holdings Ltd Agrees to Plead Guilty* (March 25 2015), available at www.justice.gov/opa/pr/schlumberger-oilfield-holdings-ltd-agrees-plead-guilty-and-pay-over-2327-million-violating-us.

12. Conclusion

Sanctions have a significant impact on the oil and gas industry globally. Those operating in this industry need to ensure that they understand the risks posed to their business by sanctions and that they have an effective corporate compliance programme to mitigate those risks. Given the powerful impact that imposing sanctions on the oil and gas sector has on a country's economy and the extent to which it is linked to the proliferation of nuclear weapons, it is likely that this sector will continue to be the target of sanctions programmes worldwide.

About the authors

Paul Aston
Partner, Holman Fenwick Willan, Singapore
paul.aston@hfw.com

Paul acts for clients in the maritime, offshore and commodity sectors (mainly oil and gas). He specialises in all aspects of shipping, transport and insurance, down and midstream oil and gas transactions, ship building and repair disputes, advising shipyards and buyers on contracts for the construction of various offshore structures (rigs/FPSO), as well as their utilisation offshore. He has particular expertise in relation to long- and short-term contracts for the supply, transportation, sale and purchase, and storage of petroleum products and LNG. He advises on commercial transactions and shipping projects.

Paul has worked in the London, Shanghai, Hong Kong and Singapore (which he founded) offices of Holman Fenwick Willan, and is presently head of the Singapore office with particular responsibility for offshore and energy. He has been appointed by the Singapore Mediation Centre as one of their associate mediators.

Mike Birchall
Manager, Bulk Liquid Survey Department,
Minton, Treharne & Davies Ltd
mike.birchall@minton.co.uk

Mike Birchall is a former tanker deck officer and cargo inspection manager and is now manager of the bulk liquid cargo survey department at Minton, Treharne & Davies Ltd (MTD). He also specialises in the technical investigation of claims and disputes relating to bulk liquids and gases, with particular experience in terminal and tanker operations, marine cargo surveys, and shortage investigations. He has conducted surveys on over 2,000 vessels worldwide and has provided technical advice on numerous mitigation operations involving complex reprocessing and/or blending operations on board tankers and in terminals. He frequently acts as an expert witness in legal disputes concerning such matters. He is the author of courses concerning many aspects of the marine shipment of oils, gases and chemicals, and has provided advice to a number of technical committees concerning the measurement and sampling of bulk liquids in terminals and on board tankers.

Simon Cook
Partner, Sullivan & Worcester UK LLP
scook@sandw.com

Simon Cook is a partner in the Trade & Export Finance Group in the London office of Sullivan & Worcester UK LLP. He has experience in a wide variety of banking and finance transactions, including, in particular, structured trade finance, trade finance, project finance, invoice discounting facilities and borrowing-base facilities in Africa, the Middle East, Asia and the CIS. His work in the structured trade area covers a range of financings acting for both lenders and borrowers, notably in the oil, telecommunications, soft commodities and metals sectors, with particular experience in Africa and the Middle East. He has worked and travelled extensively in Africa and the Middle East, having spent over three and a half years in Dubai. He has also participated in a number of structured trade finance and project finance conferences and seminars throughout Europe, the Middle East and Africa.

Jane Fitzgerald

Senior associate, Ince & Co LLP
Jane.Fitzgerald@incelaw.com

Jane is a trade and shipping specialist. She advises companies that trade, finance and transport commodities, shipowners, operators and charterers on a range of contentious and non-contentions matters.

Jane's expertise includes all aspects of physical commodity trading across a broad range of commodities. She advises her trader clients in relation to international sale and purchase, documentary credits, shipping/transportation arrangements, delivery terms and commodity finance. In particular she has crude and oil products, coal (including ScoTA), gas, LNG, grain and metals experience.

In addition to litigation in the English Courts and cross-border arbitration under all of the major international rules, Jane acts for trader clients in relation to industry specific institutional arbitration rules such as LME, GAFTA and FOSFA. Jane also regularly advises her trader clients on the non-contentious aspects of commodity transactions.

Cheryl Guan

Senior associate, Ince & Co LLP, Singapore
cheryl.guan@incelaw.com

Cheryl Guan is a Senior Associate with the Singapore office of Ince & Co LLP and specialises in corporate insurance. She advises regional and international insurer and broker clients on a range of insurance mandates across multiple lines, including marine, aviation, financial lines and casualty matters. Cheryl has extensive dispute resolution experience in contentious work, ranging from litigation to mediation and arbitration, in respect of both defence and coverage work. She also advises on non-contentious matters, including policy wordings, underwriting agreements, regulatory matters, claims management agreements and other insurance-related commercial agreements.

Denys Hickey

Barrister and Arbitrator, 39 Essex Chambers
denys.hickey@39essex.com

Denys Hickey is a barrister with 39 Essex Chambers in London, Singapore and Kuala Lumpur. He was for many years a partner with Ince & Co in London and Singapore, and has extensive experience in Asia. His work has covered all aspects of oil and gas trading, including shipping, trading, finance and competition law. He has litigated and arbitrated on behalf of oil majors, national oil companies and some of the biggest independent traders for well over 30 years, and has advised on general terms and conditions for all types of trading contract.

Catherine Jago

CEO, CJH Energy Limited
catherine@cjhenergy.co.uk

Catherine Jago has over 30 years' experience in the commercial side of the oil industry and has worked for and with many major and private companies, including BP, ICIS-LOR, Poten, Ecopetrol, Sasol, Sonangol and KPC. She has been an oil trader, price reporter, pricing analyst, management consultant, oil broker, price forecaster and oil journalist. She has given expert advice in over 100 cases involving crude oil and oil products, many of which have required an analysis of hedging strategies undertaken and the creation of hedging models. As well as offering consultancy and expert witness services in the field of oil trading, pricing, hedging and shipping matters, she acts as an arbitrator in these and in closely related fields, such as commodity trading and hedging.

Anna Koshy
Associate, Sullivan & Worcester UK LLP
akoshy@sandw.com

Anna Koshy is an associate in Sullivan & Worcester UK LLP's London office. Her practice area is banking and finance, with an emphasis on trade and export finance. She has advised a number of leading international financial institutions on a range of cross-border deals, including a variety of commodities from oil and gas to metals to agricultural products, high value pre-export finance transactions, supply chain finance, commodity repurchase transactions and risk participations. She has also advised on matters involving various trade finance instruments, as well as regulatory issues currently affecting trade finance practitioners. She was a contributor to *A Practitioner's Guide to Trade and Commodity Finance* (2012).

Hannah Laming
Partner, Peters & Peters Solicitors LLP
HLaming@petersandpeters.com

Hannah Laming is a partner in the Business Crime department of Peters & Peters Solicitors LLP, having previously been called to the Bar of England & Wales in 2001 and having acted both as an SFO prosecutor and as a member of the FSA's Wholesale Enforcement Division. She advises on criminal and regulatory matters, including economic sanctions and export controls, serious fraud, corruption, private prosecutions, internal investigations and contentious FCA issues. She has assisted individuals subject to UK, EU and/or OFAC sanctions; and regularly advises corporations about the impact of sanctions and export controls on their business in the UK and overseas; the development of appropriate policies, procedures and controls to mitigate sanctions risks; due diligence and the investigation and remediation of breaches of sanctions. She regularly speaks and writes on topics relating to international trade compliance, sanctions and export controls and was recently named as one of the world's top 100 female investigative lawyers.

John Minton
Managing director, Minton, Treharne & Davies Ltd
john.minton@minton.co.uk

John Minton is a graduate chemical engineer and is the managing director of the Minton, Treharne & Davies Group of companies (MTD), located in the UK, USA, Singapore, Australia and the Netherlands. He has forty years' experience of the technical handling of well over 4,000 refining quantity and quality claims arising from the international trade in and shipment of petroleum and petro-chemicals. His duties have included determining causation and responsibility as well as advising on mitigation (where he has developed innovative solutions). He also acts as an expert witness in related legal disputes. He has frequently advised the London insurance market on the wording of bulk oil insurance policies and has served as a member of a British Standards Institute Petroleum Technical Committee concerned with bulk petroleum measurement. In that capacity he was the UK-appointed expert to two International Standards Organisation committees dealing with such issues.

Richard Minton
Manager, Bulk Liquid Quality Department,
Minton, Treharne & Davies Ltd
richard.minton@minton.co.uk

Richard Minton is the manager of the bulk liquid cargo quality claims department at Minton, Treharne & Davies Ltd (MTD). He holds a Master's degree in Chemistry from the University of Oxford and specialises in the technical investigation of claims and disputes relating to bulk liquids, with particular experience in petroleum, petrochemical, vegetable oil and biofuel cargoes, bunker disputes, onshore power generation fuel issues and the application of forensic analytical methodologies to suspected contamination incidents. He is experienced in providing advice and recommendations concerning mitigation. He is a frequent presenter at industry seminars and is a member of a number of technical committees relating to the storage, carriage and handling of petroleum, petrochemical and biofuel cargoes.

Jasvinder Nakhwal
Partner, Peters & Peters Solicitors LLP
JNakhwal@petersandpeters.com

Jasvinder Nakhwal is a partner with Peters & Peters Solicitors LLP in London. She has in-depth experience advising corporate and individual clients on white-collar crime and related extradition, INTERPOL and mutual legal assistance treaty matters. She is recognised as a leading expert on the law of criminal cartels and is ranked in the top tier for cartel investigations in the *Chambers UK 2016* legal directory. Her experience in this field includes acting for two British Airways executives who were acquitted of allegations of price-fixing with Virgin Atlantic Airways and representing the managing director of Dunlop Oil and Marine, the first person to be charged with the criminal cartel offence in the UK. She is a contributing author to *The Law of Criminal Cartels* (2009, OUP).

Ben Olbourne
Barrister, 39 Essex Street
ben.olbourne@39essex.com

Ben Olbourne is an English barrister specialising in international commercial disputes across a range of sectors, including international trade and commodities, energy, natural resources, construction and shipping. He has extensive experience of upstream and downstream oil and gas disputes. He has appeared as counsel before courts at all levels in England and in arbitrations subject to different governing laws and under all of the leading arbitral rules. Since 2012, he has practised principally from Maxwell Chambers in Singapore. As an arbitrator, he is a Fellow of the Chartered Institute of Arbitrators and is presently or has recently been appointed in arbitrations under the ICC, LCIA, SIAC, LMAA and other rules. He is recommended in the leading UK and regional directories for commodities (*Legal 500 UK*, 2013–2015), international arbitration (Chambers & Partners *Global and Asia-Pacific Guides 2014–2016*, *Legal 500 Asia-Pacific 2016*), energy (*Legal 500 Asia-Pacific 2016*), and shipping (*Legal 500 Asia-Pacific 2016*).

Iain Sharp
Partner, Rodyk & Davidson LLP
iain.sharp@rodyk.com

Iain Sharp is an English-qualified solicitor who trained in London (including a six-month seat in-house at Shell International). He is a partner in Rodyk & Davidson's Litigation & International Arbitration Group. He has previously worked in London, Hong Kong and Singapore since 2010. His sector experience includes international trade and commodities, energy and natural resources (including oil and gas); shipping, aerospace and defence; and banking and finance.

He regularly advises clients on contracts, sale and purchase/off-take agreements, general trading terms and conditions, joint venture agreements, agency/distribution, charterparties, economic sanctions, operational and legal risk management, trade finance and letters of credit. He also conducts international arbitrations under institutional rules, including the ICC, LCIA, HKIAC, SIAC and Swiss Rules, as well as under trade association rules, including the LME, GAFTA, FOSFA and PORAM.

Stuart Shepherd
Partner, Ince & Co LLP
stuart.shepherd@incelaw.com

Stuart Shepherd joined Ince & Co in London as a trainee in 1982. He became a partner in the firm in 1990 and is now Global Head of Trade. He has specialised in shipping and commodity trading throughout his career, acting for traders in various industries, from oil and gas to commodities traded on GAFTA and FOSFA terms. He is recommended by a number of legal directories, including Chambers and Partners and the *Legal 500*.

John Simpson
Partner, Ince & Co LLP, Singapore
john.simpson@incelaw.com

John Simpson is a partner in the Singapore office of Ince & Co LLP, specialising in shipping, energy and international trade. He advises clients on charterparties, bills of lading, hedging transactions, FFAs and shipping derivatives. He also advises on the negotiation and drafting of construction and asset employment contracts relating to oil and gas services and associated disputes, and on contracts and disputes in relation to the trade of oil, bulk cargoes (including coal and iron ore) and LME metals. He acts for a diverse portfolio of clients including multinational metals and petroleum companies, major contractors to energy companies, P&I Clubs, shipowners and builders, charterers, international insurers and reinsurers and global banks operating in the Asia Pacific, East and West Africa and India. He has extensive experience of arbitration in Singapore (SIAC, SCMA, UNCITRAL and *ad hoc*), London (LMAA, LCIA, ICC and *ad hoc*), Kuala Lumpur (KLRCA) and Indonesia (BANI), as well as litigation in the English High Court.

Jonathan Tickner
Partner, Peters & Peters Solicitors LLP
JTickner@petersandpeters.com

Jonathan Tickner is a leading litigator, specialising in large and complex commercial cases including competition disputes, civil fraud, asset tracing and investigations. He has extensive experience dealing with international cases involving multi-jurisdictional emergency procedures. He has noted expertise in competition disputes, including stand-alone and follow-on private enforcement actions (eg, *Legal 500 UK 2015*).

Philip Woolfe
Barrister, Monkton Chambers
pwoolfe@monckton.com

Philip Woolfe was called to the Bar in 2004 and has practised at Monckton Chambers since 2006. He is ranked by Chambers UK as a leading junior in the fields of competition law and telecommunications. He has acted in numerous competition regulatory investigations and damages actions, including most recently in relation to electricity and gas distribution, financial markets and pharmaceuticals. He is a contributing author to Bellamy & Child, *European Union Law of Competition* (7th Edition, 2013, OUP). He was a Fellow of All Souls College, University of Oxford from 2003 to 2010. He is a committee member of COMBAR and a member of the Competition Law Association, the Bar European Group and the Procurement Lawyers' Association.

Rebecca Wright
Criminal Barrister, 23 Essex Street Chambers
rwrightlaw@gmail.com

Dr Rebecca Wright is a criminal barrister who completed a secondment with Peters & Peters Solicitors LLP in London from January to July 2015. She previously worked as an associate for Latham & Watkins LLP in Qatar, focusing principally on oil and gas matters. In addition, she spent a number of years working as an international human rights lawyer in the Middle East and Afghanistan. She has a JD from Harvard and Berkeley law schools and a DPhil from the University of Oxford.

www.ingramcontent.com/pod-product-compliance
Ingram Content Group UK Ltd.
Pitfield, Milton Keynes, MK11 3LW, UK
UKHW060930200526
471236UK00002B/9

9 781911 07803